Neural Basis of Consciousness

Advances in Consciousness Research

Advances in Consciousness Research provides a forum for scholars from different scientific disciplines and fields of knowledge who study consciousness in its multifaceted aspects. Thus the Series will include (but not be limited to) the various areas of cognitive science, including cognitive psychology, linguistics, brain science and philosophy. The orientation of the Series is toward developing new interdisciplinary and integrative approaches for the investigation, description and theory of consciousness, as well as the practical consequences of this research for the individual and society.

Series B: Research in progress. Experimental, descriptive and clinical research in consciousness.

Volume 49

Neural Basis of Consciousness
Edited by Naoyuki Osaka

Neural Basis of Consciousness

Edited by

Naoyuki Osaka

Kyoto University

John Benjamins Publishing Company

Amsterdam / Philadelphia

TM The paper used in this publication meets the minimum requirements of American National Standard for Information Sciences – Permanence of Paper for Printed Library Materials, ANSI Z39.48-1984.

Library of Congress Cataloging-in-Publication Data

Neural Basis of Consciousness / edited by Naoyuki Osaka.
p. cm. (Advances in Consciousness Research, ISSN 1381–589X ; v. 49)
Includes bibliographical references and indexes.
1. Consciousness. 2. Neurophysiology. I. Osaka, Naoyuki, 1947- II. Series.

QP411 N478 2003
153-dc21 2002035648
ISBN 90 272 5178 9 (Eur.) / 1 58811 340 X (US) (Pb; alk. paper)
ISBN 90 272 5177 0 (Eur.) / 1 58811 339 6 (US) (Hb; alk. paper)

John Benjamins Publishing Co. · P.O. Box 36224 · 1020 ME Amsterdam · The Netherlands
John Benjamins North America · P.O. Box 27519 · Philadelphia PA 19118-0519 · USA

Table of contents

Preface

This book provides a systematic examination of the neural basis of consciousness, taking as its thesis that consciousness is not a unitary function of the brain. To fully understand consciousness, therefore, we require examination of the concept of neural basis of conscious mind.

Recent advances in cognitive neuroscience made in consciousness research now make possible an understanding of the neural events that are associated with the different forms of consciousness. The book discusses the major issues related to neural basis of consciousness, the mechanisms of the different varieties of consciousness.

Research investigating the neural basis of consciousness has become a more influential stream now in 2002 as compared to 1998, when this book was conceived. It has taken several years to collect and keep up-to-date the chapters in this volume. I gratefully acknowledge continuing support from Professor Maxim I. Stamenov, series editor of the *Advances in Consciousness Research*, for his comments and suggestions. I also express my deepest thanks to the contributors of each chapter for updating the manuscript during the editorial processes.

One of the contributors, Sir John C. Eccles (who was a Nobel Prize Laurente in 1963) passed away on May 2nd, 1997 in Locarno, Switzerland, after submitting the chapter with Professor Beck. We regret his death and would like to express our thanks to his great contribution to modern consciousness research.

Naoyuki Osaka
Kyoto, July 20th, 2002

Chapter 1

Issues in neural basis of consciousness

An introduction

Naoyuki Osaka
Kyoto University, Japan

Consciousness is a most important issue for human beings and has been a central problem among philosophical debate since the modern Cartesian age. However, consciousness has also been assumed to be a mystery that was withdrawn from scientific investigation. Consciousness plays an essential role in high-level cognition, such as perception, language comprehension, self-recognition, mental operations, complex reasoning and problem solving. However, the neural basis of consciousness (NBC) is probably not yet an essentially unveiled issue. Many scientists studying consciousness consider that the evidence and theory currently have to be developed in a sustained way, and a firmer understanding of consciousness is now being accelerated by new evidence from cognitive neuroscience, cognitive psychology, neurophilosophy, neuropsychology and "new neurophysics", i.e., quantum brain dynamics. Since Descartes in the 17th century, philosophical debates on the issue of consciousness and brain relation has a long history and the theories of consciousness can be globally categorized into dualistic and monistic views of the mind and body relation. Furthermore, the current theory of consciousness introduced computational and/or non-computational views in terms of conscious information processing in the brain.

The ability to process new information while simultaneously holding onto the previous results of processing is essential for high-level conscious cognition. Current evidence from cognitive neuroscience and computational neurobiology indicate that the neural mechanism supporting consciousness may be in the central executive functions in a prefrontal cortex (PFC) in both the dorsolateral (DL) and ventrolateral (VL) PFC, intra-laminar thalamus, anterior cingulate cortex (ACC) and, orbito-frontal cortex, with coordination across these and other brain areas being related to task-dependent processes. Reflecting its central role in human cognition, the neural basis of consciousness has recently been a hot topic in the scientific study of consciousness.

Scientific study of NBC has made significant progress during the last 10 years and has reached a critical point at which detailed comparisons of different theoretical and experimental proposals are not only possible but would be tremendously beneficial for further theoretical development of the field. As the title indicates, the central rationale behind this book is to provide such a forum for comparisons of existing models and theories of consciousness in the brain. Dualistic and monistic approaches are contrasted with computational and non-computational approaches to human consciousness.

This theoretical focus reflects a current need in the field of cognitive science, cognitive neuroscience, neuropsychology, and neurophilosophy. Whereas existing theories of consciousness are quite diverse and each provide a sophisticated and well-developed account of certain aspects of the neural basis of consciousness and function, different models have different theoretical emphases and tend to leave some other aspects of consciousness relatively unspecified.

The main goal of this book is to alleviate this problem and encourage further research among different consciousness researchers by focusing on explicit, detailed comparisons of current major consciousness theories and models. More specifically, the book fulfills this goal by asking leading consciousness theorists in the world to address the neural basis of consciousness, attention, and qualia that have been guiding recent research in this area.

Consciousness, how it works and how it arises within the neuronal network in the human brain has been investigated using sophisticated psychological techniques coupled with functional neuroimaging technology such as functional magnetic resonance imaging (fMRI). Although in the light of modern fMRI, the relation between the brain and mind appears as an unresolved issue. According to the philosopher Chalmers (1995), the brain and mind issue relates to "the hard problem" of qualia in the conscious brain, while the binding issues of neuroscience find this a rather "easy problem". Thus the "easy problem" of the philosopher becomes the "hard problem" for the neuroscientist. This is a serious issue, since the relation between brain and mind can never be fully understood until the nature of consciousness is explicable in terms of subjective quality of the sense called qualia. How does qualia work in the brain? Qualia is sometimes argued as something that will never be resolved by science, while some theorists represent qualia using recursive neural nets (The Churchlands, this volume).

In an emerging new biological science of consciousness, multidisciplinary scientists including, cognitive psychologists, cognitive neuroscientists, neuropsychologists, neurophilosophers, mathematicians, and quantum physicists are now facing the same unique issue although they view the issue from different multidisciplinary angles. While some scientists regard consciousness as an essential product of neuronal activities in the brain, others regard it as a by-product of quantum superposition in the orchestrated reduction within a microtuble of the axonal dendrite (Hameroff & Woolf, this volume) or even due to quantum probabilistic exocytosis of the apical dendrite of the brain (Beck & Eccles, this volume).

There is no way out of these mysteries of consciousness except by way of multidisciplinary approaches in the context of a scientific investigations. In the present volume, we propose and follow up these scientific approaches and present the reader with the leading edge of the neural basis of consciousness to solve the mystery of consciousness.

The first step concerns the nature of the neural basis of consciousness and the control and regulation of consciousness. What are the mechanisms that constrain the neural basis of consciousness and how is the information inside consciousness controlled and regulated? Further, what is the role of the neural basis of consciousness in complex cognitive activities. How is consciousness involved in complex cognitive activities, such as perception, memory, attention, language, and self-consciousness? What is the relationship between consciousness and working memory? Are they different entities? Or is consciousness simply an activated portion of working memory? In the chapters on neuronal modeling, brain models of the neural mind are proposed.

In chapters that include working memory, working memory-based neural models are proposed. Working memory and Global Workspace (GW) theory are both cognitive models that are now increasingly shaped by brain evidence. Both aim to explain significant aspects of perception, inner speech, mental imagery, and the like. And both come from long-standing research traditions: Working memory from short-term memory experiments and GW theory from cognitive architectures (Baars, this volume). GW theory was developed to account for the role of conscious elements in cognition. According to Baars, consciousness is not an empty side-effect of brain functioning but it is king of the hill: All active mental processes make use of it. In this theory, conscious WM elements recruit unconscious resources to carry out their jobs.

Using working memory as a basis, the chapter by Osaka (2001) proposed a cognitive psychobiological model of consciousness by extending active working memory architecture, in which the brain's dorsolateral prefrontal cortex (DLPFC) and anterior cingulate cortex (ACC) play a critical role in integrating various high-level information subsets of working memory. The model proposes distributed network-structured central executive systems in DLPFC (self-monitoring) and ACC, i.e., attentional control and information update/switching, with memory subsystems in the occipital, temporal, and parietal cortices. Thus, working memory has proven to be a key concept to understand neural basis of higher brain function including consciousness and to describe the global cognitive architecture of three-layered structured consciousness. The working mind works on the basis of capacity-constrained and goal-directed central executive in working memory.

The storage and processing function in temporarily co-activated brain areas distributed in parallel over the brain are the essential neural basis of the conscious mind. Thus, to put the neural puzzle together in an effort to a general theory of the neural basis of consciousness, the chapter by Cotterill argues that the neural correlates of consciousness should be considered in his explanation of the manner in which the underlying brain circuits could support totally covert simulation of the body's muscular interaction with the environment. Thus, he tried to make his theory in detail by reviewing of his previous beliefs on the subject.

The chapter by Libet investigates how neural events in the brain are related to the production of conscious and unconscious mental functions. Electrical stimulation of somatosensory sites with intracranial electrodes in awake human subjects indicated that awareness requires that cerebral activities last up to 500 msec. Sensory experience is thus delayed, but subjectively the timing is antedated to the earliest input to the cortex. In a voluntary act, brain activity was found to precede awareness of conscious intention by about 400 msec. Voluntary acts are thus initiated unconsciously, but conscious function can prevent the action by veto. The transition from an unconscious detection to sensory awareness also requires an increase in duration of cerebral activation. This "time-on" model has important implications. Subjective experience may be viewed as an attribute of a "conscious mental field (CMF)". The CMF emerges from, but is phenomenologically distinct from brain activity.

In chapters that include neuronal psychophysics, a neural model of visual perception is proposed and attention is contrasted with consciousness in visual awareness.

The reason for perceptual multistability experienced when viewing various figures, most likely lies in the brain's physical organization: an organization that imposes several constraints on the processing of visual information. Why is it that our visual system fails to lock onto one aspect of an ambiguous figure?

What are the neural events that underlie such changes? Are there neurons in the visual pathways the activity of which reflects visual awareness of the stimulus? The chapter by Logothetis et al. describes some combined psychophysical and physiological experiments that were motivated by these questions. Specifically, they report on experiments in which neural activity in the early visual cortex and in the inferior temporal cortex of the monkeys was studied, while the animals experienced binocular rivalry. Their results provide us with new evidence not only on the neural mechanisms of binocular rivalry, one example of multistable perception, but also on the neural processes underlying image segmentation and perceptual grouping.

What is the relationship between consciousness and attention? Do these terms refer to the same construct? William James (1890) considered attentional processing separate from consciousness and that both were required for mindfulness. Lately, however, the assumption has been that attention and consciousness are intimately related, if not identical. In particular, the recent excitement over the possibility that a thalamocortical circuit (most likely among the reticular nucleus, parietal cortex, and the prefrontal lobes) is responsible for conscious experience confuses attentional processing with phenomenal experience. The chapter by Hardcastle discusses why James's original concept is correct. We can be conscious of some low level aspects of complex stimuli on which we are not focusing attention, and some classical attentional phenomena, such as orienting to abrupt changes in our environment, can occur without conscious awareness. Though our attentional processing may be a diffuse process that encompasses large regions of the cortex, our projection systems, and various thalamic areas, we need to look at its site – and what it is acting upon – to locate our phenomenal experiences. Implicit priming data and lesion studies indicate that consciousness is best conceived of as residing in the parietal cortex, where attention acts on a subset of our conscious perceptions and thoughts.

In the chapters on neural philosophy, consciousness is argued in terms of cognitive philosophy. Broad spectrum philosophical resistance to physicalist accounts of conscious awareness has condensed around a single and clearly identified line of argument. In the chapter by the Churchlands it is argued that philosophical analysis and criticism of that line has also begun to crystallize. The nature of that criticism coheres with certain theoretical ideas from cognitive neuroscience that attempt to address both the existence and the contents of consciousness. Furthermore, experimental evidence has recently begun to emerge that will serve both to constrain and to inspire such theorizing. In this chapter, the Churchlands attempt to summarize the situation.

In the chapters that include quantum computing of consciousness, quantum mind in the brain are proposed as the scientific basis of consciousness and quantum computing in microtubles is also discussed.

Beck and Eccles argue the so-called "psychon" hypothesis incorporating the quantum process in an apical dendrite. A new and intriguing view of the relation between the brain and consciousness arises, however, if quantum processes play a decisive role in brain activity. The quantum state reduction, or selection of amplitudes, resting deeply in the principles of the theory, offers a possible doorway for a new logic, quantum logic, with its unpredictability of a single event. Brain activity consists of a constant firing of neural cells, regulated by synaptic switches which establish connections between neurons. Conscious action, e.g. intention, is a dynamical process which forms temporal patterns in some areas of the brain. They discuss how synaptic activity in the form of exocytosis of transmitter molecules can be regulated effectively by a quantum trigger based on electron transfer processes in the synaptic membrane. Conscious action is hereby essentially related to quantum state reduction.

The chapter by Hameroff and Woolf discusses quantum computing in microtubles and argues related problems. According to their view, conventional models portray consciousness as "emerging" from complex neurophysiological activities among neurons, however such emergence theories lack specific thresholds or rationale and generate no testable predictions. On the other hand the "Penrose-Hameroff Orchestrated Objective Reduction" ("Orch OR") model portrays consciousness as deriving from brain processes which access proto-conscious properties at the fundamental level of the universe, precisely specifies the threshold at which consciousness occurs, and generates testable predictions (Hameroff-Woolf, this volume). They argue that Orch OR depends on quantum computation in microtubules within neurons, tuned by microtubule-associated protein-2 (MAP-2). To accommodate Orch OR they outline a neural circuit in the cerebral cortex centered around dendrites of the large pyramidal cells, which are enriched with microtubules and MAP-2. Thus, Hameroff and Woolf insist that new approaches are needed. The Penrose-Hameroff model suggests that quantum superpositions and Penrose's objective reductions occur in microtubules in groups of brain neurons and glia interconnected by gap junctions. The proposed microtubule quantum states and cycles of self-collapse are isolated by actin gelation, orchestrated by microtubule-associated proteins, and coupled to neural-level activity, e.g. coherent 40 Hz oscillation. Consciousness may involve neurobiological processes extending downward within neurons to the level of the cytoskeleton, and accessing fundamental experience at the most basic level of reality. Thus, Orch OR model can account for conscious experience, associative memory and volitional choice.

In the last chapter, Scott argues quantum theories of the mind. Over the past year or two, discussions of the roles that quantum mechanics might or might not play in the theory of consciousness have become increasingly sharp. On one side of this debate stand conventional neuroscientists who assert that brain science must look to the neuron for understanding, and on the other side are physicists, suggesting that the rules of quantum theory might influence the dynamics of mind. He presents a middle path between the vagueness of theories on quantum consciousness and the sterility of the neural network approach. How is this possible? Both sides of the current debates, in his opinion, give insufficient consideration to the explanatory power of classical nonlinear theory. Properly appreciated, the ramifications of intricate nonlinear systems obviate the need for turning to quantum theory as a source of the mysteries of mind. In the same token, this approach rescues the neural network theory from the shallows of narrowly conceived functionalism.

Since the volume is based mainly on a biological approach, brain processes are quite rigidly defined in terms of the role they play in meeting the individual's needs. Biological approaches to consciousness are critically concerned with functional descriptions.

These propositions explain the subjective and objective aspects of conscious experience in terms of the underlying brain functioning.

References

Chalmers, D. J. (1995). Facing up to the problem of consciousness. *Journal of Conciousness Studies, 2*, 200–219.

James, W. (1890). *Principles of psychology*. New York: Dover.

Osaka, N. (1997). In the theatre of working memory of the brain. *Journal of Consciousness Studies, 4*, 329-331.

Part I

Neuronal modeling

Chapter 2

Working Memory requires conscious processes, not vice versa*

A Global Workspace account

Bernard J. Baars
The Neurosciences Institute, San Diego

Introduction

Working Memory and Global Workspace theory are cognitive models that are now increasingly shaped by brain evidence (Baddeley 1998; Baars 2002). Both have been worked out in considerable detail. Both aim to explain significant aspects of inner speech, purposeful mental imagery, immediate memory, and the like. And both come from long-standing research traditions: Working Memory from short-term memory experiments and Global Workspace theory from cognitive architectures.

In spite of these similarities, they differ in interesting ways. Global Workspace (GW) theory was developed to account for the role of conscious elements in cognition. In GW theory, contrary to skeptics, consciousness is not an empty side-effect of brain functioning. On the contrary, it is king of the hill: All active mental processes make use of it. That is presumably why all active components of Working Memory (WM) are conscious to one degree or another – including inner speech, the currently rehearsed item, input, novel visual imagery, and decisions to rehearse, recall, and report. In GW theory, conscious WM elements recruit unconscious resources to carry out their jobs. This is demonstrated in a large-scale GW simulation that models classical WM functions (Baars & Franklin, in preparation).

Working Memory theorists are now exploring consciousness. Some suggest that consciousness requires WM (Baddeley 1992; Andrade 2001). Yet it is more likely that WM functions emerged from consciousness: first, because all active WM operations involve consciousness, not vice versa; second, because sensory consciousness appears to date back to early mammals, more than 100 million years ago, while WM depends upon hominid language and prefrontal functions, perhaps a few million years old (Courtney et al. 1998); and third, because GW theory can account for basic WM phemonena such as rehearsal, but not the other way around.

In recent years Global Workspace theory has gained a considerable amount of support from philosophers like Daniel Dennett (2001), neuroscientists like Edelman and Tononi (2000), and cognitive scientists like Chris Frith, Dehaene, et al. (2001), Kanwisher (2001) and Franklin (2002). Many seem to agree on the usefulness of a global workspace capability in the brain for understanding consciousness. A number of recent brain discoveries also support Global Workspace theory (Baars 1998, 2002).

There are striking similarities between Working Memory (WM) and Global Workspace theory (GW) (Baddeley 1992, 1998; Baars 1988, 1997, 1998, 2002; Andrade 2001). But the familiar Working Memory functions all depend on consciousness and voluntary control, which are essentially left unexplained in WM theory. GW aims to understand them.

Consider Working Memory rehearsal, for example. In classic WM studies people are given a set of unrelated items to keep in immediate memory, either numbers, words, or nonsense syllables. Of the traditional 7 plus or minus two items one can typically keep in mind using rehearsal, at any given moment only one or two are conscious. The others are unconscious. But the currently conscious item can be manipulated: It can be deliberately rehearsed, or ignored; it can be perceived, recalled, and reported; it can often be recoded either verbally or visually. That is not the case for unconscious items. The only way to manipulate unconscious items is to first recall them to consciousness. Thus consciousness appears to be the final common path of all standard WM operations. If we add the role of conscious instructions to perform any of these operations, and the fact that the goal of carrying out instruction must be conscious (reportable) as well, the centrality of consciousness can no longer be ignored.

Figure 1 shows a simplified WM model as of 2001. The small symbols added to each component indicate whether it is unconcious (dark circle), conscious (white circle), or fringe conscious (red star). For example, items in episodic Long Term Memory (LTM) are unconscious. They cannot be reported in the standard model without being brought to consciousness first, via the episodic buffer (or perhaps the phonological loop or visual sketchpad). By contrast, Baddeley (2000) defines the output of the episodic buffer as conscious.

A model of Working Memory (Andrade 2001)

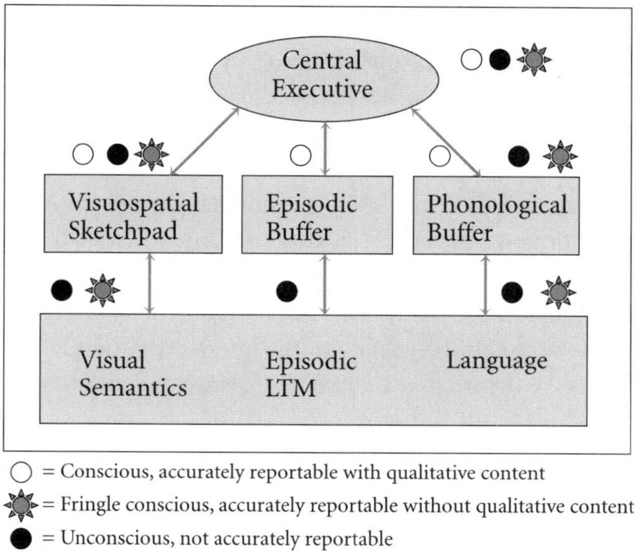

Figure 1. Working Memory elements seem to include conscious, unconscious, or fringe conscious aspects.

Operational definitions

Consciousness is defined operationally by accurate report; unconsciousness by an inability to report accurately; and fringe consciousness by accurate report without reported sensory qualities, as in the famous tip-of-the-tongue state. These simple operational measures have been used for decades in a great variety of experimental studies. They can be used to test all hypotheses suggested here.

Notice that most WM components appear to be mixtures of conscious, unconscious and fringe conscious events. For example, the Central Executive appears to have some fringe conscious elements, as indicated by judgments about ourselves. If one is asked whether one is like a psychotic as murderer, most people would deny it. But such denial is a fringe-conscious judgment. That is, it can be highly accurate, but it does not possess sensory qualities like the color red, object properties, figure-ground contrast, clear spatial and temporal boundaries, and the like. Fringe conscious judgments may involve prefrontal cortex (Baars 2002b).

Most Central Executive functions are surely unconscious. That is, they are not accurately reportable at all. There is a large experimental literature in social psychology showing how inaccurate people can be about themselves, suggesting unconscious processes. Likewise, brain damage to orbitofrontal cortex impacts executive self-control, as in the famous case of Phineas Gage. But during normal functioning people cannot report the details of self-control; it seems to be largely unconscious. Thus while executive functions constantly interact with conscious contents, they seem to be largely unconscious in their details.

WM studies emerged from an Anglo-American intellectual tradition that ruled out consciousness for many years (Baars 1986). For that reason, WM theorists have come to the problem only in the last decade (Baddeley 1992, 2000; Andrade 2001). GW theory may be able to provide some useful ideas for them.

The basics of Global Workspace theory

Global Workspace theory is a simple cognitive architecture that has been developed to account qualitatively for a large set of matched pairs of conscious and unconscious processes (Baars 1983, 1988, 1993, 1997). Such matched contrastive pairs of phenomena can be either psychological or neural. Psychological phenomena include binocular rivalry, visual backward masking, priming, automaticity with practice, and selective attention. Neural examples include the thalamocortical core (Edelman & Tononi 2000), specific visual deficits, and blindsight.

In theoretical terms, GW theory has only three constructs: (1) unconscious networks that are specialized for numerous tasks; (2) contexts, which are unconscious networks that shape conscious contents; and (3) the global workspace, a capacity for global distribution and integration for information. All of cortex may be said to consist of specialized and largerly unconscious cells, columns, blobs, and networks. A good example of contextual networks are the egocentric and allocentric maps in parietal cortex, which are apparently unconscious in themselves, but which are necessary to maintain a full conscious visual field. The brain damage called right parietal neglect involves a deficit in these parietal maps, which causes a perceived collapse of the left side of visible space. Neuronal candidates for a global workspace capacity include sensory cortices, prefrontal cortex, and perhaps some nuclei of the thalamus.

Figure 2 shows the three constructs and how they are used in GW diagrams (from Baars 1988). Such diagrams may be used to show the flow of information in complex functions, like WM recall.

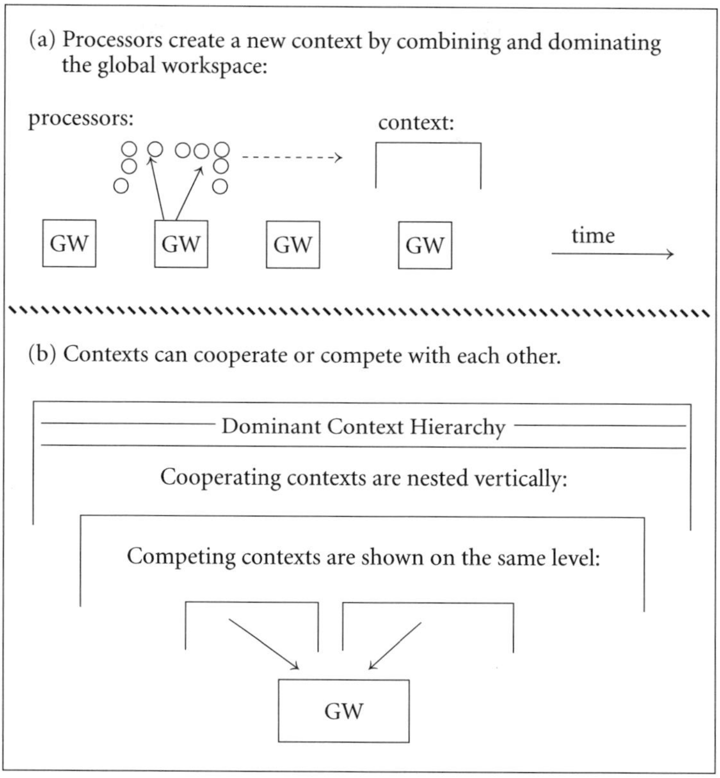

Figure 2. Specialized unconscious networks may combine to shape conscious contents contextually. An example is the unconscious egocentric and allocentric maps of the posterior parietal cortex, which shape the conscious visual field (from Baars 1988).

Notice that executive functions in GW theory are shown as a Dominant Context Hierarchy, that is, a stack of contexts, corresponding to unconscious expectations and intentions that shape conscious contents. Since executive functions are essential in WM operations, the Dominant Context Hierarchy or goal stack is an important part of the following discussion.

Like other cognitive architectures, GW theory may be thought of as a theater of mental functioning. The theater metaphor is too simple, but it offers a useful first approximation. Consciousness in the metaphor resembles a bright spot on the stage of immediate memory, directed there by a spotlight of attention, under executive guidance. The rest of the theater is dark and unconscious.

This approach leads to specific neural hypotheses. For sensory consciousness the bright spot on stage is likely to require the corresponding sensory projection areas of the cortex. Sensory consciousness in different modalities may be mutually inhibitory, within approximately 100-ms time cycles. Sensory cortex can be activated internally as well as externally, resulting in the "internal senses" of conscious inner speech and imagery. Once a conscious sensory content is established, it is distributed widely to a decentralized "audience" of expert networks sitting in the darkened theater, presumably using corticocortical and corticothalamic fibers. This is the primary functional role of consciousness: to allow a theater architecture to operate in the brain, in order to integrate, provide access, and coordinate the functioning of very large numbers of specialized networks that otherwise operate autonomously (Mountcastle 1978). All the elements of GW theory have reasonable brain interpretations, allowing us to generate a set of specific, testable brain hypotheses about consciousness and its many roles in the brain. Some of these ideas have now received considerable empirical support (Baars 2002).

Inner speech, imagery, and Working Memory. Both auditory and visual consciousness can be activated endogenously. Inner speech is a particularly important source of conscious auditory-phonemic events, and visual imagery is especially useful for spatial orientation and problem solving. The areas of the left hemisphere involved in outer speech are now known to be involved in inner speech as well (Paulesu, Frith, & Frackowiak 1993). Likewise, mental imagery is known to involve most of visual cortex (Kosslyn 1988). Inner speech and the "visuospatial sketchpad" are often viewed as the two basic components of cognitive Working Memory (WM) (Baddeley 1992).

Likewise, internally generated somatosensory imagery may reflect emotional and motivational processes, including internally generated feelings of pain, pleasure, hope, fear, sadness, etc. Such internal sensations may communicate to other parts of the brain via global broadcasting. Prefrontal executive systems may not have direct access to action control. Rather, they may work by evoking motivational imagery, broadcast from the visual cortex, to control relevant parts of motor cortex, thereby generating appropriate actions. Parts of the brain that play a role in emotion may also be triggered by global broadcasting of conscious contents from sensory cortices and insular cortex. For example, it is established that the midbrain amygdalae are needed to recognize facial expressions of fear and anger from the visual system (Le Doux 1996). Ultimately such areas, working together, shape actions controlled by frontal cortex and subcortical automatic systems like the basal ganglia.

Thus many cortical areas work together to transform goals and emotions into actions (Baars 1987; Baars, Fehling, McGovern, & LaPolla 1996). Since there are many spatial maps throughout the brain, the trade language of the brain may consist of activated maps, paced by temporal oscillations in the alpha to gamma spectrum. Such oscillations may coordinate the activity of multiple sensory, body space, and external spatial maps.

The attentional spotlight. The sensory "bright spot" of consciousness involves a selective attention system (the theater spotlight), under dual control of frontal executive cortex and automatic interrupt control from areas such as the brain stem, pain systems, and emotional centers like the amygdala. It is these attentional interrupt systems that allows significant stimuli to "break through" into consciousness in a selective listening task, when the name is spoken in the unconscious channel. Posner (1992) and colleagues have found such a cortical system for executive visual attention.

Context vs. content of sensory experience. A conscious sensory "bright spot" requires the interaction of sensory analyzers and contextual systems. In vision sensory content seems to be produced by the ventral visual pathway, while contextual systems in the dorsal pathway define a spatial domain within which the sensory event is defined. As pointed out above, parietal cortex is known to include allocentric and egocentric spatial maps, which are not themselves objects of consciousness, but which are required to shape every conscious visual event.

There is a major difference between the symptoms of dysfunctional *content* systems such as the visual ventral stream, compared to disordered context systems. In the case of lesioned content, the subject can generally notice a missing part of normal experience; but for damage to context, the system of experiential expectations is itself damaged, so that one no longer knows what to expect, and hence what is missing. This may be why parietal neglect is so often accompanied by anosognosia, a massive but very specific loss of knowledge about one's body space (Bisiach & Geminiani 1991; Schacter 1990). Patients suffering from parietal anosognosia may reject their own body limbs, or see themselves with three hands. Such specific loss of contextual body information is not accompanied by a loss of general intelligence or knowledge about the world. Damage to such contextual knowledge systems may be quite specific.

Self-systems. Cortical activation by a seen object may not be enough to generate subjective consciousness. The activated visual information may need to be sent to self-systems, which serve to maintain constancy of an inner framework across different perceptual situations. When we walk from room to room in a building, we must maintain a complex and multi-leveled organization that can be viewed in Global Workspace theory as a higher-level context. Major goals, for example, are not changed when we walk from room to room, but conscious perceptual experiences do change. Gazzaniga (1996) has found a number of conditions under which split-brain patients encounter conflict between right and left hemisphere executive and perceptual functions. He has proposed the existence of a "narrative self" in the left frontal cortex, based on split-brain patients who are clearly using speech in the left hemisphere to talk to themselves, sometimes try to force the right hemisphere to obey its commands. When that proves impossible, the left hemisphere will often rationalize or reinterpret the sequence of events so as to repair its understanding of the interhemispheric conflict.

Analogous repairs of reality are observed in other forms of brain damage, such as parietal neglect. They commonly occur whenever humans are confronted with major, unexpected life changes. The left-hemisphere narrative interpreter may be considered as a higher-level context system that maintains expectations and intentions across many specific situations. While the narrative stream itself is conscious, it is shaped by unconscious contextual executive influences.

If we consider Gazzaniga's narrative interpreter of the left hemisphere to be one kind of self-system in the brain, it must receive its own flow of sensory input. Visual input from one half of the field may be integrated in one visual hemisphere, as described above, under retinotopic control from area V1. But once it comes together in visual cortex, it needs to be conveyed to frontal areas on the left side of the brain, in order to inform the narrative interpreter of the current state of perceptual affairs. The left prefrontal self system then applies a host of criteria to the input, such as "Did I intend this result?", "Is it consistent with my current and long-term goals?", "If not, can I reinterpret the input to make sense in my running account of reality?".

It is possible that the right hemisphere has a parallel system that does not speak, internally or externally, but that may be more able to deal with anomalies via irony, jokes, and other emotionally sophisticated strategies. The evidence appears to be good that the right prefrontal cortex can understand such figurative uses of language, while the left does not.

Full consciousness may not exist without the participation of such prefrontal self systems.

New brain evidence for global distribution of conscious contents

On the basis of psychological evidence, GW theory predicted that conscious contents are widely distributed in the brain (Baars 1988, 1997). Today that case is supported by a sizable body of brain evidence, at least for sensory consciousness (Baars 2002). For example, Dehaene and colleagues showed that backward-masked visual words evoked brain activity confined to the well-known visual word recognition areas of cortex (Dehaene et al. 2001). Identical conscious words triggered higher levels of activity in these areas, but more importantly, they evoked far more widely distributed activity in parietal and prefrontal cortex. That result has now been replicated a dozen times, using different brain imaging techniques, different experimental comparisons between conscious and unconscious input (e.g. binocular rivalry, inattentional blindness, neglect, extinction), and different modalities (audition, pain, vision, and sensorimotor tasks). In all cases conscious sensory input evoked far wider and more intense brain activity than identical unconscious input.

These findings support the general claim that conscious stimuli mobilize large areas of cortex, presumably to distribute information about the stimuli. This idea is essential to the following discussion. If consciousness serves to mobilize many unconscious specialized networks, the active elements of WM that always need to be conscious – input, recall, rehearsal, inner speech, visual imagery and report – may be widely distributed in order to recruit specific unconscious functions needed to carry out those tasks.

If that is true, *WM activities are always mediated by conscious elements.* That is the basic claim made here.

We are now ready to sketch out how a GW model would handle a mental operation like recall from WM.

Explaining recall from Working Memory

We are now ready to give a GW account of the conscious components of WM. We will focus here on WM recall, the voluntary retrieval of an item from the half dozen unrelated items that are typically held in WM. This has been standard in experiments over the years. In the literature WM recall is usually treated as a single operation. And yet it requires a complex set of simpler operations that require consciousness.

GW theory aims to account for both consciousness and voluntary control. In WM models conscious/voluntary aspects include item retrieval, report,

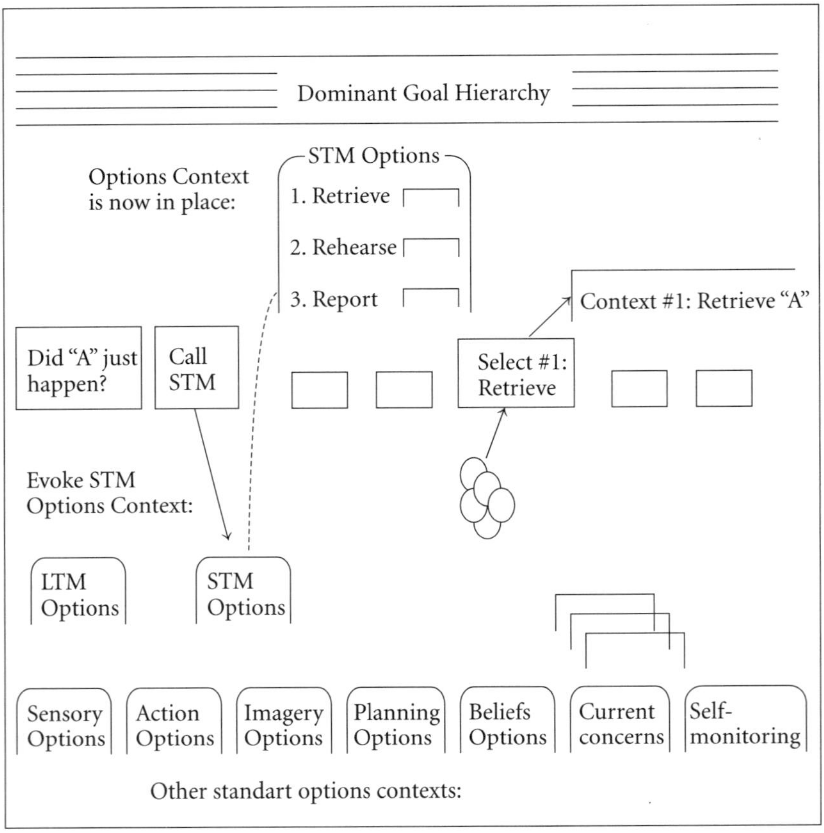

Figure 3. A Global Workspace account of recalling an event from Working Memory. (Note that Working Memory was called Short Term Memory or STM in 1988.)

agreement to carry out the instructions, and the like. Perception of WM input is conscious but not voluntary. Further, WM functions involve voluntary efforts to control selective attention, defined in GW theory as the selection of conscious contents (see Frith, in press). Thus in trying to understand WM recall, we must have a clear concept of consciousness, voluntary control, and selective attention, both voluntary and involuntary.

In this approach voluntary recall involves an implicit "options context", that is, a representation of alternative WM items that could be recalled on the basis of availability and other features. In a typical computer one can find the contents of memory by calling up a "directory", a list of files. From the directory one can choose one or another file, and engage in reasoning about the different options: one file may lack the desired information; another may have it in an awkward format; a third may be so long that the desired information will be hard to find, and so on. Seeing the directory permits one to engage in conscious reasoning about the options. The same necessity and solution seem useful in the nervous system. It would be nice to have rapid conscious access to the alternatives which one could recall from WM. That is the role of the options context.

The options context is guided by goals, but it is not rigidly constrained by invariable goals, as automatic functions appear to be. Rather, it is contingent on multiple influences, including contexts and preceding conscious events. That is why it is useful for the act of voluntary item retrieval to be conscious. In GW terms, that is why it makes sense to "broadcast" the choices between items to many knowledge sources, which may then interact to bias the process of item selection as needed.

Specialized networks can "vote" for the various options presented, and the winning option could evoke the appropriate effectors and subgoals by ideomotor control, as in Baars (1988) (Models 4 and 5). Thus we seem to have all the makings for a mental directory already at hand, as shown in Figure 3 (shown as a rounded frame with the options listed inside). An Options Context shows the available options for future conscious contents. Once the Options Context dominates the global workspace it presents a "menu" or "directory" of possible conscious contents; the most relevant one is chosen by a decision process, which may include "votes" from specialized networks, as well as from the Goal Hierarchy; the winning option then elicits another action context by global broadcasting.

Which came first, consciousness or WM? Evolutionary considerations

A number of neurobiologists suggest that sensory consciousness emerged with early mammals or transitional reptiles (therapsids), who evolved a large thalamocortical complex (e.g. Edelman & Tononi 2000). WM involves two endogenous kinds of sensory consciousness, the phonological loop (for inner speech), and the visual sketchpad (for voluntary visual imagery). The phonological loop seems to use the classic speech regions of the left hemisphere, Broca's and Wernicke's areas (Paulesu et al. 1993). Wernicke's area is located near auditory cortex, and in the case of inner speech it seems to support internalized auditory speech perception. Likewise, visual imagery appears to involve visual cortex, especially when it is vividly conscious (Kreiman et al. 2000).

Working Memory involves the purposeful use of such endogenous sensory consciousness. It is not clear when the purposeful use of visual imagery and inner speech evolved, but it plausibly did so *after* the major growth of prefrontal cortex and speech areas associated with hominid evolution – that is, in the last few million years. Prefrontal cortex is key to voluntary goals in humans, and speech is of course the basis of the phonological loop involved in mental rehearsal.

In contrast, spontaneous imagery may have appeared much earlier – such as the visual image of a lion evoked by the sound of a lion's roar. Such spontaneous imagery may be common among mammalian prey animals, while predators may have spontaneous visual images of prey that can be smelled but not seen. In humans, spontaneous speech and imagery occur even when subjects are asked not to engage in any deliberate thought (Mazoyer et al. 2002), and it seems likely that spontaneous mental events like this may occur quite early in mammalian evolution. Certainly the dream state, characterized by rich conscious visual imagery, evolved with early mammals.

Thus both external and endogenous sensory capacities may have evolved millions of years *before* the hominid capacities needed for Working Memory. These considerations suggest that endogenous sensory consciousness long precedes WM, and that the distinctive hominid development of WM depends on voluntary control of these pre-existing functions. This is especially true of language and its role in voluntary control of inner speech and purposeful visual imagery.

This is not to deny, of course, that conscious contents and WM functions are constantly interacting in the human brain. Evolutionary primacy does not imply functional one-directional causation in a brain process.

Summary and conclusions

Consciousness seems to be needed for all the active elements of Working Memory. What role does it play in WM tasks? Global Workspace theory suggests that consciousness enables multiple networks to cooperate and compete in solving problems, such as retrieval of specific items from immediate memory.

The overall function of consciousness is to provide widespread access, which in turn may serve functions of coordination and control. Consciousness is the gateway to the brain, enabling control even of single neurons and whole neuronal populations (Baars 1988). None of these control functions become directly conscious, of course, but conscious feedback seems required to recruit control by prefrontal networks. In the metaphor of the theater, it is as if each specialized audience member can decide locally whether or not to be driven by input from the bright spot on stage. Executive functions – the director behind the scenes – is also largely unconscious, often using the actor in the spotlight on the stage of working memory capacity to recruit and trigger specific functions.

The brain is a radically distributed organ; yet there are executive functions that can influence any part of the nervous system via the "broadcasting" function of conscious experience. WM is one of many functions that can be controlled by conscious events.

Notes

* This work was supported by the Neurosciences Institute and the Neurosciences Research Foundation, which is gratefully acknowledged.

1. Dennett and Kinsbourne have criticized a concept called a "Cartesian Theater" (Dennett & Kinsbourne 1990). Although Global Workspace theory can be viewed metaphorically as a theater, it is not a Cartesian theater. It lacks a point center, for example. GW theory comes from a long tradition of working computational architectures, unlike the Cartesian Theater, which is an unworkable *reductio ad absurdum*. Dennett himself now agrees with the notion of a neuronal global workspace (see Dennett 2001).

References

Anderson, J. R. (1983). *The architecture of cognition.* Cambridge, MA: Harvard University Press.

Andrade, (2001). *Working Memory in perspective.* London: Psychology Press.

Baars, B. J. (1983). Conscious contents provide the nervous system with coherent, global information. In R. Davidson, G. Schwartz, & D. Shapiro (Eds.), *Consciousness and self-regulation* (pp. 45–76). New York: Plenum Press.

Baars, B. J. (1986). *The cognitive revolution in psychology.* NY: Guilford Press.

Baars, B. J. (1987). What is conscious in the control of action? A modern ideomotor theory of voluntary control. In D. Gorfein & R. Hoffmann (Eds.), *Learning and Memory: The Ebbinghaus Centennial Conference.* NY: L. Erlbaum.

Baars, B. J. (1988). *A cognitive theory of consciousness.* New York: Cambridge University Press.

Baars, B. J. (1993). How does a serial, integrated and very limited stream of consciousness emerge from a nervous system that is mostly unconscious, distributed, and of enormous capacity? In G. R. Bock & J. Marsh (Eds.), *CIBA Symposium on Experimental and Theoretical Studies of Consciousness* (pp. 282–290). London: John Wiley and Sons.

Baars, B. J. (1997). *In the theater of consciousness: The workspace of the mind.* Oxford University Press.

Baars, B. J. (2002). The conscious access hypothesis: Origins and recent evidence. *Trends in Cognitive Sciences, 6* (1), 47–53.

Baars, B. J. (2002b). A prefrontal hypothesis of fringe consciousness. *Psyche* (in press).

Baars, B. J., & Newman, J. (1994). A neurobiological interpretation of the Global Workspace Theory of consciousness. In A. Revonsuo & M. Kamppinen (Eds.), *Conscousness in philosophy and cognitive neuroscience.* Hillsdale, NJ: Erlbaum.

Baars, B. J., & McGovern, K. (1996). Cognitive views of consciousness: What are the facts? How can we think about them? In M. Velmans (Ed.), *The science of consciousness: Tutorial reviews.* London: Routledge.

Baars, B. J., Fehling, M., McGovern, K., & LaPolla, M. (1996). Competition for a conscious global workspace leads to coherent, flexible action. In J. Cohen & J. Schooler (Eds.), *Scientific approaches to consciousness: The 25th Carnegie Symposium in Cognition.* Hillsdale, NJ: Erlbaum.

Baddeley, A. (1992). Consciousness and Working Memory. *Consciousness & Cognition, 1* (1), 3–6.

Baddeley, A. (1998). Recent developments in Working Memory. *Current Opinion in Neurobiology, 8,* 234–238.

Bisiach, E., & Geminiani, G. (1991). Anosognosia related to hemiplegia and hemianopia. In George P. Prigatano & Daniel L. Schacter (Eds.), *Awareness of deficit after brain injury: Clinical and theoretical issues.* NY: Oxford University Press.

Bogen, J. E. (1995). On the neurophysiology of consciousness: I. An overview. *Consciousness & Cognition, 4* (1).

Chase, M. H. (Ed.). (1974). *Operant control of brain activity.* Los Angeles: University of California Press.

Courtney, S. M., Petit, L., Haxby, J. V., & Ungerleider, L. G. (1998). The role of prefrontal cortex in working memory: Examining the contents of consciousness. *Phil. Trans. Royal Society London (B)*, 1819–1928.

Crick, F. H. C. (1984). Function of the Thalamic Reticular Complex: The searchlight hypothesis. *Proceedings of the National Academy of Sciences USA 81* 4586-93, July.

Crick, F. H. C., & Koch, C. (1990). Towards a neurobiological theory of consciousness, *Seminars in the Neurosciences, 2,* 263–275.

Damasio A. R. (1989). Time-locked multiregional retroactivation: A systems-level proposal for the neural stubstrates of recall and recognition. *Cognition, 33,* 25–62.

Dehaene, S., & Naccache, L. (2001). Towards a cognitive neuroscience of consciousness: Basic evidence and a workspace framework. *Cognition, 79,* 1–37.

Dennett, D. C. (2001). Are we explaining consciousness yet? *Cognition, 79,* 221–237.

Dennett, Daniel C., & Kinsbourne, M. (1992). Time and the observer: The where and when of consciousness in the brain. *Brain and Behavioral Sciences, 15,* 183–247.

Edelman, G. (1989). *The remembered present: A biological theory of consciousness.* NY: Basic Books.

Edelman, G. M., & Tononi, G. (2000). *A universe of consciousness.* NY: Basic Books.

Franklin, S. (2002). What IDA says about conscious and unconscious language processing and verbal reports. Sixth Annual Meeting of the Association for the Scientific Study of Consciousness. Barcelona, Spain.

Baars, B. J., & Franklin, S. (under review). How conscious events may interact with Working Memory: Using IDA, a large-scale model of Global Workspace theory.

Gazzaniga, M. S. (Ed.). (1996). Consciousness and the cerebral hemispheres. In *The Cognitive Neurosciences*. Cambridge, MA: Bradford/MIT Press.

Geschwind, N. (1979). Specializations of the human brain. *Scientific American, 241* (3), 180–201.

Greenwald, A. J. (1992). Is the unconscious simple? *American Psychologist.*

Kanwisher, N. (2001). Neural events and perceptual awareness. *Cognition, 79*, 89–113.

Kosslyn, S. M. (1988). Aspects of a cognitive neuroscience of mental imagery. *Science, 240*, 1621–1626.

Kreiman, G., Koch, C., & Fried, I. (2000). Imagery neurons in the visual brain. *Nature, 408*, 357–361.

LeDoux, J. (1996). *The emotional brain.* NY: Simon & Schuster.

Leopold, D. A., & Logothetis, N. K. (1996). Activity changes in early visual cortex reflect monkey's percepts during binocular rivalry. *Nature, 379*, 549–553.

Llinás, R., & Ribary, U. (1992). Rostrocaudal scan in human brain: A global characteristic of the 40-Hz response during sensory input. In E. Basar & T. Bullock (Eds.), *Induced rhythms in the brain* (pp. 147–154). Boston, MA: Birkhäuser.

Logothetis, N. K., & Schall, J. D. (1989). Neuronal correlates of subjective visual experience. *Science, 245*, 761–763.

Logothetis, N. K. (1999). Vision: a window on consciousness. *Scientific American*, Nov., *281*(5), 69–75.

Luria, A. R. (1980). *Higher cortical functions in man* (2nd ed.). NY: Basic. Russian language edition 1969.

Mazoyer, B., Zago, L., Mellet, E., Bricogne, S., Etard, O., Houde, O., Crivello, F., Joliot, M., Petit, L., & Tzourio-Mazoyer, N. (2001). Cortical networks for working memory and executive functions sustain the conscious resting state in man. *Brain Research Bulletin, 54* (3), 287–298.

Mountcastle, V. B. (1978). An organizing principle for cerebral function: The unit module and the distributed system. In G. M. Edelman & V. B. Mountcastle (Eds.), *The mindful brain.* Cambridge, MA: The MIT Press.

Newell, A., & Simon, H. A. (1972). *Human problem solving.* Englewood Cliffs, NJ: Prentice-Hall.

Newell, A. (1990). *Unified theories of cognition.* Cambridge, MA, Harvard University Press.

Newman, J., & Baars, B. J. (1993). A neural attentional model for access to consciousness: A Global Workspace perspective. *Concepts in Neuroscience, 4* (2), 255–290.

Paulesu, E., Frith, D., Frackowiak, R. S. J. (1993). The neural correlates of the verbal component of working memory. *Nature, 362*, 342–345.

Posner, M. I. (1992). Attention as a cognitive and neural system. *Current Directions in Psychological Science, 11*, 11–14.

Rosch, E., & Lloyd, B. (1978). *Cognition and categorization.* NJ: Erlbaum.

Schacter, D. L. (1990). Toward a cognitive neuropsychology of awareness: Implicit knowledge and anosognosia. *Journal of clinical and experimental neuropsychology, 12* (1), 155–178.

Scheibel, A. B. (1980). Anatomical and physiological substrates of arousal In J. A. Hobson & M. A. Brazier (Eds.), *The Reticular Formation revisited.* NY: Raven Press.
Shallice, T. (1976). The Dominant Action System: An information-processing approach to consciousness. In K. S. Pope & J. L. Singer (Eds.), (1978), *The stream of consciousness: Scientific investigations into the flow of experience.* NY: Plenum.
Sperry, R. W. (1966). Brain bisection and mechanisms of consciousness. In John C. Eccles (Ed.), *Brain and conscious experience.* New York: Springer-Verlag.
Spritzer, H., Desimone, R., & Moran, J. (1988). Increased attention enhances both behavioral and neuronal performance. *Science, 240*, 338–340.
Weiskrantz, L. (1986). *Blindsight.* Oxford, England: Oxford University Press.
Zeki, S. (1993). *A vision of the brain.* London: Blackwell Scientific.

Chapter 3

Working memory-based consciousness

An individual difference approach

Naoyuki Osaka
Kyoto University, Japan

Introduction

Consciousness, including awareness in a broader sense, has been a critical issue in philosophical debates since the modern Cartesian age began in the 17th century. Consciousness has been assumed to be a mystery which had been withdrawn from the scientific investigation of modern behaviorism. However, recent findings on higher brain function in cognitive neuroscience has reactivated consciousness research and the Cartesian mind/body problem. This investigation is shared by multidisciplinary scientists, such as neuropsychologists, computational neuroscientists, neurobiologists, cognitive psychologists, cognitive philosophers, "new" physicists, and information scientists.

Cognitive neuroscience has revealed that consciousness, generating from complex neural networks dynamically connecting various areas of the brain, plays an essential role in high-level cognition, such as perception, language comprehension, self-recognition, mental operations, reasoning and problem solving. However, the neural basis of consciousness (NBC) is likely not as yet an essentially unveiled issue. Many scientists studying consciousness believe that the evidence and theory currently have to be built in a sustained way, and the development of a firmer understanding of consciousness is now accelerating due to new evidence from cognitive neuroscience and cognitive psychology. Furthermore, working memory has recently been noted as an essential contributor to NBC: working memory plays an essential role in developing a firmer understanding of consciousness and attention (Osaka 1997, 1998). Working memory refers to the process whereby information is temporarily maintained and concurrently processed in capacity-constrained active memory for use in ongoing goal-directed achievement. Further, working memory system is assumed to consist of verbal as well as spatial short-term storage and central executive processes that work on these storage systems (Baddeley 1986).

The first step, in the present chapter, concerns the nature of the neural basis of working memory and the control and regulation of working memory. The second step concerns recent fMRI evidence based on the resource sharing model of working memory involving a capacity-constrained concurrent activated system (Just & Carpenter 1992).

Working memory in awareness

What is the relationship between consciousness and working memory? Working memory, of what it works and how it arises in the neuronal network in the human brain has been revealed using fMRI (functional Magnetic Resonance Imaging), PET (Positron Emission Tomography), and MEG (Magnetoencephalography). At present, the relation between consciousness and working memory is not clear enough. However, we propose the layered model in Figure 1 to show their interrelation. Figure 1 (left panel) shows, a three-layered model of consciousness: arousal-, awareness-, and recursive- consciousness from bottom to top. These layers supposed to interact with each other bidirectionally and arousal, awareness, and the recursive layer each corresponds to vigilance, perceptual-motor- and self-consciousness. Vigilance is based on biologically-driven awareness, and perceptual-motor-awareness is based on perception of environment as well as motor control of the goal-directed behavior, while self-consciousness is based on high-level awareness. In the right panel, the corresponding layered model of working memory is shown: the neurotransmitter-, awareness-related working memory, and high-level working memory system, respectively. In this section, we will review awareness working memory and its binding mechanism specifically in the visual working memory system, which has been extensively investigated recently. Then we will move to high-level verbal working memory and related fMRI evidence in the next section.

Recent studies in monkeys and humans have emphasized the important role of the prefrontal cortex (Fuster 1995). Neurobiological segregation of processing is an important principle of the neural basis of visual awareness and perception. Ungerleider and Mishkin (1982) found, even in modularity, largely separate streams of processing: A "ventral stream" through the inferior temporal cortex processes information about features that identify objects, such

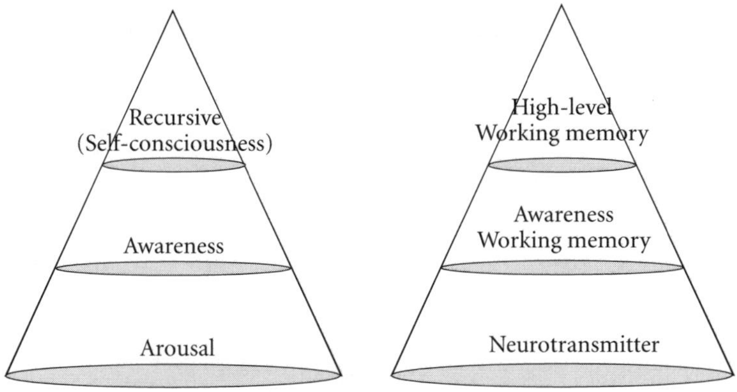

Figure 1. Layered model of consciousness and corresponding working memory model (Osaka 1998).

as shape, color and shape (object, or "what" stream), and a "dorsal stream" through the posterior parietal cortex processes information about location and spatial relations among object (spatial, or "where" stream). Such dual stream in visual awareness raises the question of how and where information about object identity is bound with information about object location. One brain area that may play a critical role in binding is the prefrontal cortex, which receives inputs from all the sensory systems in the brain. Furthermore, a major contribution of the prefrontal cortex to high-level cognition is the active maintenance and temporary storage of information, a process known as working memory (Baddeley 1996). Prefrontal neurons that contribute to working memory are candidates for binding diverse streams such as the what- and where-streams. Information from these two streams is received by separate areas of the prefrontal cortex, the dorsolateral prefrontal cortex (DLPFC; BA(Brodmann Area)46/9) and the ventrolateral prefrontal cortex (VLPFC; BA12), respectively (Ungerleider 1995). However, there are interconnections between these areas that could bring what and where together (Rao et al. 1997). To investigate whether information from these two streams is bound by individual prefrontal neurons, Rao et al. (1997) using a task in which what and where are used together and found, using a monkey, that what and where signals could be integrated through interconnections between DLPFC and VLPFC, through converging projections from the parietal and temporal cortex on the frontal cortex, through cross-talk in the visual cortex. Thus, these prefrontal neurons may contribute to the linking of object information with spatial information, needed to guide behavior. In functional brain imaging studies, mostly sim-

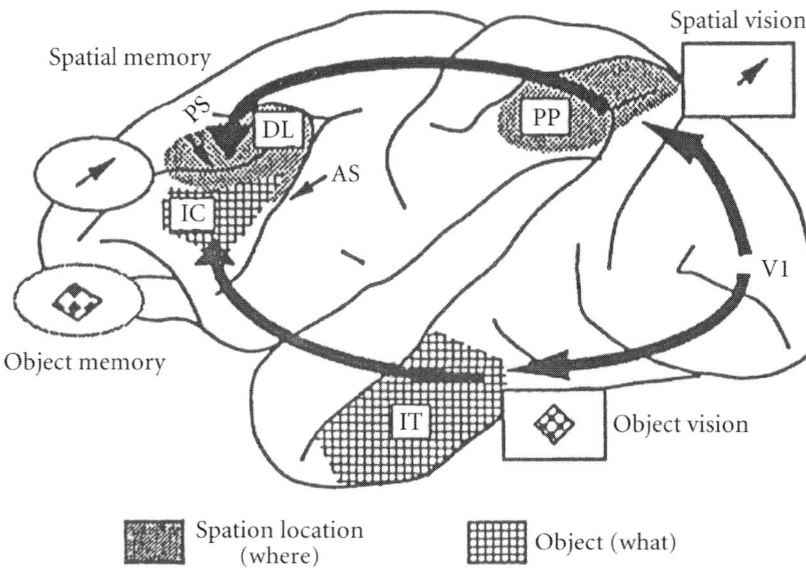

Figure 2. Two streams of visual working memory in monkey. Spatial vision (dorsal or "where" stream) and object vision (ventral or "what" stream) based on domain-specific visual working memory hypothesis (Wilson et al. 1993). PP: Posterior Parietal, IT: Inferior Temporal, DL: Dorsolateral Prefrontal, IC: Inferior Convexity, PS: Principal Sulcus, AS: Arcuate Sulcus.

ilar evidence was reported for the visual awareness working memory. It has been claimed that VLPFC plays a modality-specific (domain-specific) role in form, face and color working memory (Courtney et al. 1996); while according to two-level hypothesis suggested by Petrides (1994, 1996), DLPFC is involved in second-order manipulation and monitoring of both spatial and verbal information in working memory (Rushworth et al. 1997; Petrides 1996). These are so-called process-specific working memory and involves high-level verbal working memory which incorporated self-ordered task and n-back tasks (Petrides 1996). Courtney et al. (1998) found an area specialized for spatial working memory in human prefrontal cortex using fMRI, and an area in the superior frontal sulcus that is specialized for spatial working memory was identified. Current evidences from these neuroimaging studies suggest that what and where information is bound together in the prefrontal cortex and the DLPFC is assumed to have at least a two-stage function: a modality-specific spatial working memory in its lower stage and non-spatial verbal working memory in its higher stage with monitoring function. As Figure 2 shows, the two streams of monkey visual working memory are integrated in the prefrontal cortex (Wilson et al. 1993). Thus, the prefrontal cortex is assumed to be a central executive of visual awareness.

High-level verbal working memory

Visual working memory is the neural basis of visual awareness which brings coherent perception and understanding of the outer world. But how working memory is involved in complex cognitive activities, such as active memory for language processing, comprehension, and self-consciousness? Is high-level consciousness like thinking simply an activated portion of verbal working memory? The key procedure to answer this question is developing a new method of measuring the capacity of verbal working memory.

Daneman and Carpenter (1980) have developed a test paradigm to measure individual executive working memory capacity using working memory span tests, which involve a reading span test (RST), and a listening span test (LST), that share the common property that these tasks require for concurrent processing and storage (such as listening for comprehension) and short-term maintenance (such as remembering the last word in a each sentence). Unlike a single-task test such as digit span (or word span), working memory span test has prove to be a reliable predictor of performance on a wide variety of verbal working memory measures. Thus, a dual-task-based working memory span test is found to be a sensitive measure of working memory capacity and efficiency (Osaka & Osaka 1994). Several psychological studies on working memory have been reported using RST or its listening equivalent LST (Daneman & Carpenter 1980; Osaka & Osaka 1994). These measuring procedures are based on a resource sharing model in which working memory is essentially capacity-constrained and have individual differences in nature (Just & Carpenter 1992). An executive function such as multiple task coordination accompanied by attention shifting, self-monitoring and temporary verbal memory updating are most critical roles for high-level cognition and conscious processes and assumed to be subserved by the prefrontal cortex (Fuster 1995, 1997). Verbal working memory shares the function to the temporary storage with/and processing of goal-directed information, and is assumed to involve specific temporal storages and executive functions, such as attention-based coordination, planning, that operate on the information in verbal working memory. The executive part of the working memory is likely to be comprised of a number of distinct stages. Thus, functional brain neuroimaging is useful in separating these stages to the extent that these stages critically rely on distinct neural substrates (D'Esposito et al. 1995), as we described in the previous visual working memory section. We here focus on the neural substrates of dual-task coordination during the concurrent performance of verbal working memory tasks which is closely related to high-level conscious processing including language and self-monitoring.

To test the neural mechanisms of verbal working memory in connection with attentional coordination, we employed fMRI while subjects performed a working memory LST designed specifically to evaluate executive function of the verbal working memory (Osaka et al. 1999). To test the hypothesis of individual differences in high-level conscious processing, we compared the brain area activated by the LST task in high- and low-working memory capacity subjects. Subjects were categorized into high-(HSS) and low-span subject groups (LSS) by working memory storage capacity prior to the experiments: High verbal working memory capacity subjects having LST scores above 4.0 and low verbal working memory capacity subjects having scores below 2.5 were chosen as high and low working memory subjects, respectively (on the validity for this criteria, see Daneman & Carpenter 1980; Osaka & Osaka 1994, in detail). LST is the critical test to measure the effective verbal working memory capacity, which has been developed in English (Just & Carpenter 1992) as well as in Japanese and other languages (Osaka & Osaka 1992; Osaka & Osaka 1994; Osaka 2000). These tests showed high correlation with sentence comprehension scores connected with high-level verbal working memory. We previously showed that the higher the working memory capacity, the higher the sentence comprehension due to efficient use of capacity-limited working memory for temporary storage and processing. Thus, the score is assumed to be a critical indicator of sentence comprehension and other aspects of high-level consciousness including focusing function of the central executive (Osaka et al. 2002). Using magnetoencephalography, we (Osaka et al. 1999) previously found peak alpha frequency shifted toward high when LST becomes difficult (high load working memory task) suggesting that 10 Hz band oscillation is critical for working memory function. Thus, LST yields a psychometrically robust and valid measure of both executive and storage/processing working memory capacity for individuals. However, the neural substrates of such a task have not been systematically explored in terms of high- and low-span subject comparison.

In the fMRI experiment, we used eight subjects from high and low each (Osaka et al. 2002). We employed a specialized sound-receiver system to present LST sentences during the experimental session. As Figure 3 shows, subjects performed three different types of trials in the fMRI scanner (Siemens

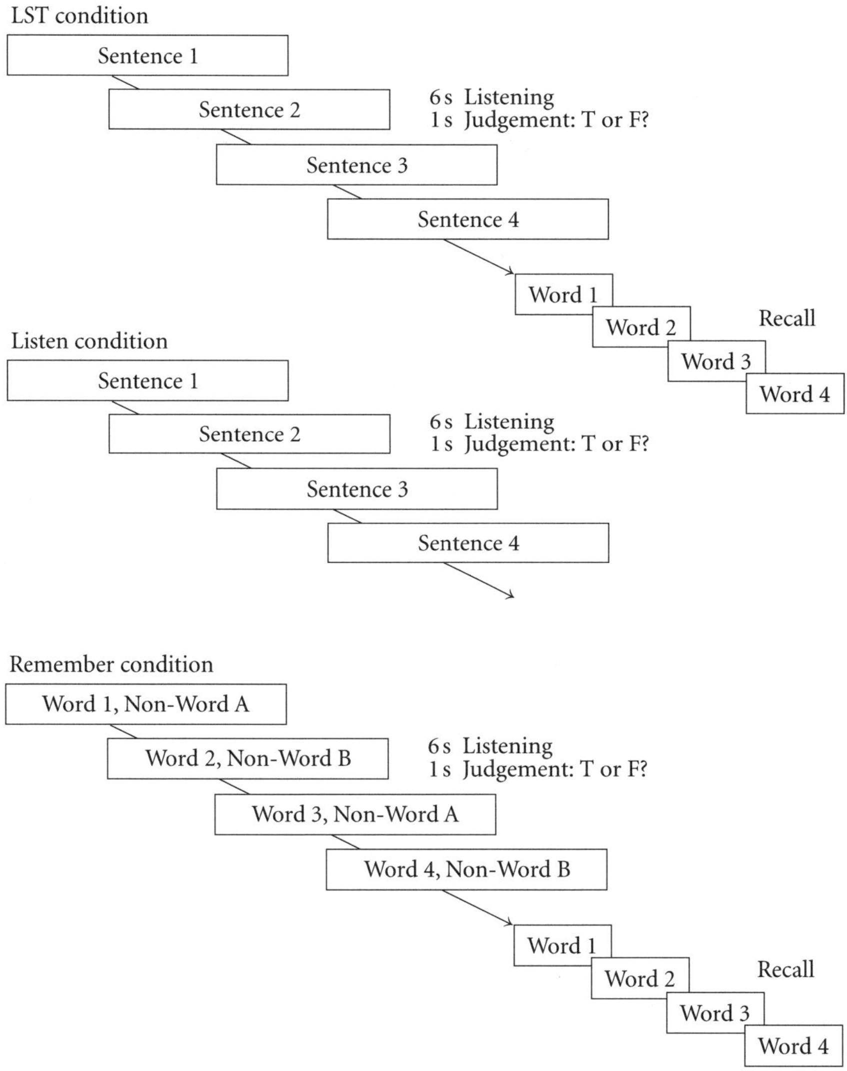

Figure 3. Time course of LST condition. Figure shows one block of LST condition which consisted on four sentences. Each sentence lasted for 6 seconds and inter-sentence-intervals was 1 second. At the end of the block, the subjects recalled the four first word of each sentence.

Magnetron Vision Imager, 1.5 Tesla whole-body type): LST, its two component tasks, and a control condition. Each trial had the same basic structure, with LST, judgment processing, and recall phases. The subject performed dual tasks based on LST, component sentence judgment task, and component word recall task, either separately or concurrently.

On judge trials, subjects evaluated four consecutive sentences as true or false by pressing a button. Subjects pressed the button if the sentence was false (e.g., "snow falls in the summer"). On recall trials, subjects recalled four words. On LST trials, subjects evaluated sentences and simultaneously recalled the target word at the beginning of each sentence. A single scan consisted of a six task blocks and seven intermittent resting blocks: One task block consisted of six sec stimulus presentation interval with a one sec inter-stimulus interval, thus requiring a total of 28 sec. On judge and LST trials, judgment of the sentence meaning was made during the inter-stimulus period. On LST and word trials, recall was required immediately after the block terminated.

Head motion was minimized using a forehead strap and functional neuroimages were obtained, using a gradient-echo-planner sequence, under the condition parameters TR = 4000 ms, TE = 60 ms, FA = 90 deg, FOV = 20 × 20 cm. We took 16 slices of image with 6mm thick with 1.2 mm gaps. T1-weighted scans were acquired for anatomical coregistration at the same locations as the functional images.

After processing with MATLAB (Mathworks Inc.), we analyzed the images using Random Effect Model from SPM99 statistical software package (Wellcome Dapartment of Cognitive Neurology, UCL, UK).We adjusted the image into EPI template(standard brain model) and applied 6 mm Gaussian filter.

The percentages of correct responses were 88.3, 96.3, and 96.7 for recall, judgment, and LST, respectively (high-group) and 79.1, 95.9, and 85.4 for recall, judgment, and LST, respectively (low group). Scores from the high-group were significantly higher than those from low-group.

The results showed that the dual task performance (LST) activated auditory cortex (Brodmann Area(BA)41) and neighboring language area (LA) such as supra-temporal gyrus including supramarginal area, Wernicke's area, and Broca's area than performance of either component task in isolation. Furthermore, as Figure 4 and 5 show, a dual task significantly activated both left dorsolateral prefrontal cortex (DLPFC: BA9). The activated areas are listed in Table 1. Moreover, interestingly enough, we found a critical activated area in the anterior cingulate cortex (ACC: BA32) for HSS and not for LSS. These findings support a resource model of working memory executive functions in the prefrontal cortex, in which verbal working memory's central executive processes are shared both by left DLPFC and ACC in a distributed network-based manner. The activated ACC area has rich connections with attention allocation brain areas like the parietal as well as the prefrontal cortex. Thus, it is

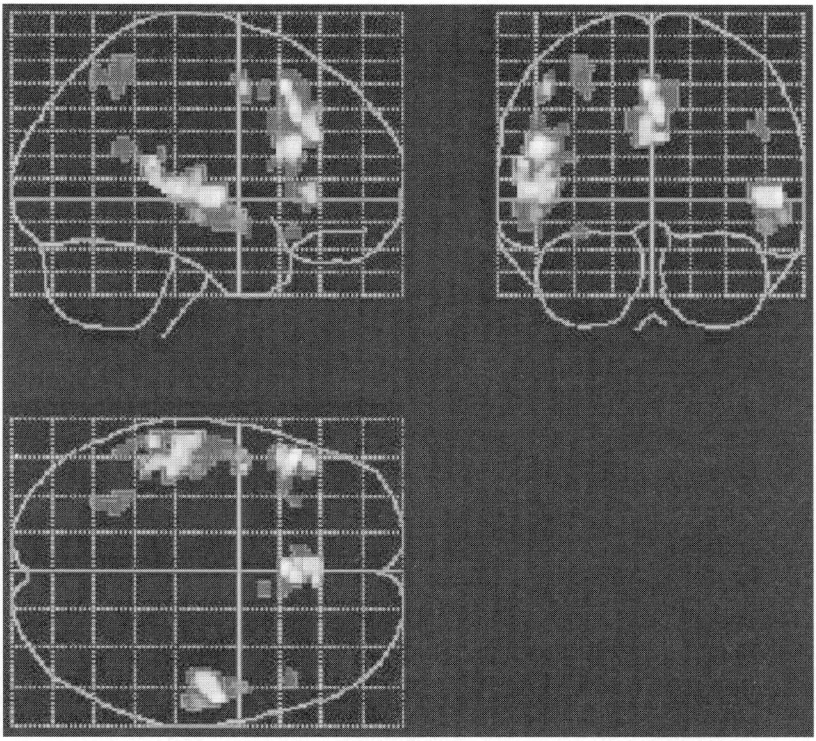

Figure 4. Panels show the activated areas in LST condition averaged across eight HSS. Activated areas on sagittal, coronal, and axial planes of the standard glass brain images (left side of the coronal and upper side of axial image each shows left hemisphere) (Osaka et al. 2002).

likely that ACC shares resource-limited attentional control system co-working with the thalamus and other parietal cortical areas. The ACC was more active when responding to incongruent stimuli (MacDonald III et al. 2000). Furthermore, activated left DLPFC appears to play an important role in the effective control of dual task processing: activation maintenance for temporarily memorized words and false/true judgment of sentence meaning during the dual task session. For example, if the subject hears a sentence like "snow falls in the summer" (with target word "snow" should be memorized and recalled after four consecutive sentences), the subject has to press a button (because it is false) and also has to keep "snow" in temporary storage so that the subject can report it after the session over (LST's dual task). However, interestingly

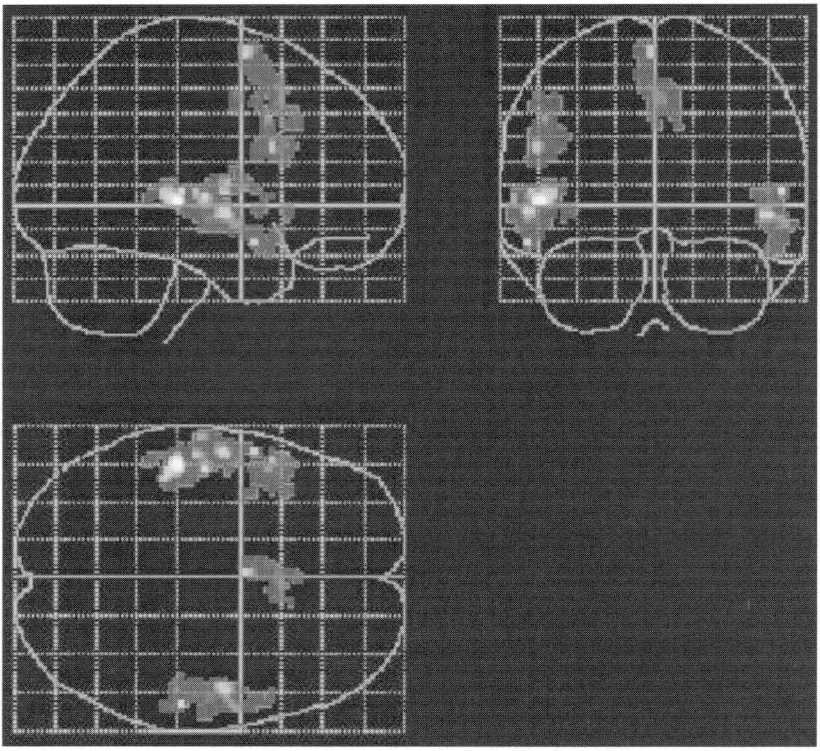

Figure 5. Panels show the activated areas in LST condition averaged across eight LSS. Activated areas on sagittal, coronal, and axial planes of the standard glass brain images (left side of the coronal and upper side of axial image each shows left hemisphere) (Osaka et al. 2002).

enough, low-group showed no activation in ACC and weaker activation over the frontal region.

This evidence suggests the individual difference in intelligence-based high-level consciousness and therefore individual differences in verbal working memory-based consciousness. Performing two tasks concurrently instead of one requires additional mental resources in the brain. Resources may be recruited from new areas specialized for dual task-specific processes, such as attention-demanding task coordination, that are not involved by either single component task. Alternately, the recruitment of additional resources may be manifested as enhanced activation in the same brain areas that subserve performance of the component tasks. The present fMRI experiment was designed to ask whether performing two tasks instead of one recruits novel dual task-

Table 1. Brain areas activated for HSS and LSS. Talairach coordinates, Z-scores, and voxel values are shown with corresponding Brodmann area (BA).

Brain region	Brodmann area	Remember Coordinates x	Remember y	Remember z	Remember Z score	Remember voxel	Listen Coordinates x	Listen y	Listen z	Listen Z score	Listen voxel	LST Coordinates x	LST y	LST z	LST Z score	LST voxel
HSS																
Frontal																
L frontal gyms	45/9						–48	24	20	3.98	56	–50	24	24	4.81	114
	45											–48	32	2	3.85	28
	44/9	–52	16	30	3.89	15										
L cingulate cortex	32						–2	34	30	3.51	11					
R cingulate cortex	32						2	24	42	4.01	93	2	24	40	4.67	173
Temporal																
L superior temporal gyrus	22	–52	–24	6	4.08	72	–54	–24	4	4.42	225	–54	–26	4	4.20	197
R superior temporal gyrus	22	56	–28	10	5.24	10	56	–10	2	4.14	134	56	–10	2	4.55	113
R middle temporal gyms	21											58	–18	–8	3.46	12
R Sylvian fissure	42	58	–16	10	4.75	43										
LSS																
Frontal																
L frontal gyms	45						–54	26	0	4.48	26					
	44/9	–54	14	30	3.73	16						–52	14	26	3.84	43
R cingulate cortex	32											2	18	46	3.56	51
Temporal																
L superior temporal gyrus	22						–64	–30	12	4.63	20	–50	–28	2	5.14	228
L middle temporal gyms	21	–54	–6	0	4.16	53	–56	6	14	3.84	20					
	21	–62	–14	0	3.83	55										
L Sylvian fissure	42						–48	–16	4	4.72	114					
R superior temporal gyrus	22	60	–24	10	4.16	73	56	–26	8	3.85	41	56	–26	6	4.09	18
R superior temporal sulcus	22											50	–6	–4	4.01	67
R middle temporal gyrus	21						58	0	–10	4.15	40					
	21						52	–18	–10	3.52	17					

specific areas or leads to increased activation in the areas recruited by component tasks. The results showed that dual task-specific areas in the same area, such as BA41 and neighboring LA including supra-temporal gyrus including the supramarginal, Wernicke's, and Broca's areas, were weakly activated. Therefore, we could conclude that high-level consciousness as revealed by high-level verbal working memory has individual differences for complex cognitive processing. Furthermore, we found the activated area for subjects with higher working memory capacity was significantly larger than that for subjects with lower working memory.

In order to examine the group differences in more detail, we compared activation differences between LST and Listen conditions in each group. In both LST and Listen conditions, the activation was found in the three main regions: left superior temporal region, left PFC and ACC. We compared the LST with Listen condition, because we found no significant activation in ACC in Remember condition.

In order to compare the activation difference between Listen and LST conditions, we measured the mean percentage changes in fMRI signal at the most activated voxels within each of the three regions of interest in each condition. Figure 6 shows percentage changes in fMRI signal in each region.

In ACC, the signal increase differed between HSS and LSS. In HSS, a significantly greater activation was found in LST than in Listen condition [$t(7) = 4.28$, $p < .01$]. The signal increase was not significantly different in the ACC for the LSS.

In the PFC, the signal increased to a greater extent in LST than in Listen condition in both HSS and LSS [$t(7) = 4.68$, $p < .01$ in HSS and $t(7) = 2.82$, $p < .05$ in LSS]. Signal increase in the left superior temporal region was approximately equal in LST and Listen conditions for both HSS and LSS groups [$t(7) = 2.01$, $p < .10$ in HSS and $t(7) = 2.02$, $p < .10$ in LSS].

The present fMRI study showed that main activation areas appeared in three regions while the subjects were engaged in the LST: temporal, PFC and ACC areas (Figure 7). These results suggest that the neural substrates of verbal working memory involve interconnections among these areas.

The first region of the network system is around the Sylvian fissure; particularly, the superior temporal gyrus near Wernicke's area (LA: language area). Just et al. (1996) found an increase of activation in these areas when the sentence comprehension became difficult. These areas are modality-specific processing system, where both verbal computation such as semantic processing and sentence comprehension are processed.

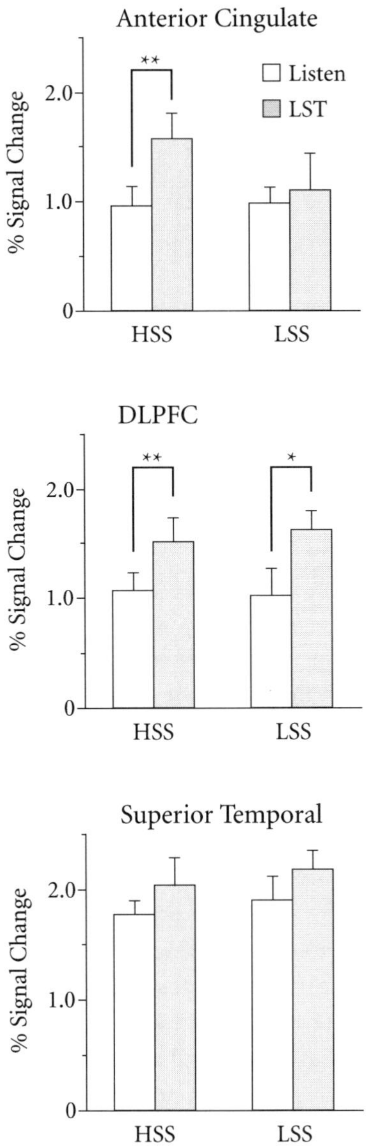

Figure 6. The mean % signal change in ACC, PFC and superior temporal cortex under LST and Listen conditions. The average of HSS is shown on the left and that of LSS on the right. Error bars represent the standard error of means. * shows there was significant difference between LST and Listen condition (**$p < .01$, *$p < .05$).

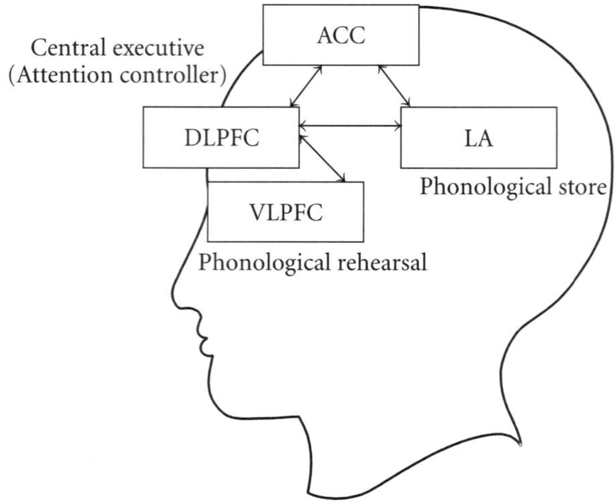

Figure 7. Functional relation among ACC (anterior cingulate cortex), DLPFC (dorsolateral prefrontal cortex), VLPFC (ventrolateral prefrontal cortex), and LA (language area). ACC and DLPFC work as the central executive (attention controller) and VLPFC and LA work as subsystem under control of the central executive.

The second region of activated network is prefrontal cortex (PFC). The activation in PFC was significantly more increased while subjects performed the LST than while they listened to a sentence for both groups. The increased activation in the frontal region confirmed the previous reports on working memory demands (Bunge et al. 2000; D'Esposito et al. 1995). D'Esposito et al. (1995) found that DLPFC activation increased only during a dual task and not during single tasks. Moreover, Bunge et al. (2000) found increased activation in this region in an RST condition that required the subjects to both read and maintain. It is conceivable that the increase in PFC was caused by dividing attentional resources between the two tasks. During the performance of LST, two different functions were concurrently executed, very similar to a dual task; listening to a few sentences on one hand and maintaining a few words on the other, both competing with each other for the capacity of working memory. Therefore, the subjects had to divide their attentional resources between both tasks, because the attentional resources are limited (Baddeley 1986; Daneman & Carpenter 1980).

It has been suggested that the functions or processes being measured during the span task is that of working memory system similar to the central executive (Baddeley 1992; Just & Carpenter 1992). It has also been suggested that the central executive is responsible for the allocation and coordination of attentional resources to the language area (Baddeley 1996). In this view, the resource allocation in LST would be controlled by the central executive system. Thus, the increased activation in the PFC in the present results most likely represents the attention control system of the central executive.

Moreover, the increase in PFC was seen both in HSS and LSS groups. In a recent study using PET, Smith et al. (2001) found activation in left PFC during operation span task only in poor performers but not in good performers. We found the increase in PFC activation in both HSS and LSS groups, which was not consistent with their results. However, this may be due to task differences between the LST used in our study and the operation span task used by Smith et al. (2001), in which task switching was required. In our LST, subjects were unlikely to use a task switching strategy between listening to a sentence and maintaining a word because the first word of each sentence was important for comprehending the sentence. They had to keep activating the first word as they parsed the entire sentence in order to judge whether the sentence was semantically true or false. Thus, listening to the sentence and maintaining the word in LST were not independent like the arithmetic and word maintainence processes in the operation span task of Smith et al. (2001). In comparison, the subjects in our study needed considerably more control of attention for both tasks making the measurable differences between the groups small.

The third region of interest is the ACC. In ACC, some researchers found activation while subjects performed a working memory task (Bunge et al. 2000; Raichle 1993). Raichle (1993) found ACC activation only in the good performers, indicating some possibility of the strategic difference between good and poor performers. In their experiment, the word maintenance was required, while in our experiment, parallel control for both maintaining words and sentence comprehension (dual task) was demanded. Therefore, ACC activation in our experiment was caused not only strategic manipulation but also the attention control such as dividing attention and monitoring the task performance.

It has been proposed that the anterior part of cingulate cortex should be characterized as "executive" in function, and the posterior parts characterized as "evaluative" (Vogt et al. 1992). Furthermore, the dorsal site of ACC is thought to be involved in cognitive activity, whereas the ventral site is an emotional division (Bush et al. 1998; Bush et al. 2000). In the present experiment, the activation was found in the dorsal site of ACC, and this is in agreement with these previous findings. Figure 8 shows possible model on neural basis of HSS and LSS: The line with arrows indicates strong interconnections between two areas while the dotted line with arrows indicates weak interconnections between two areas. ACC has strongly connected with both DLPFC and LA for

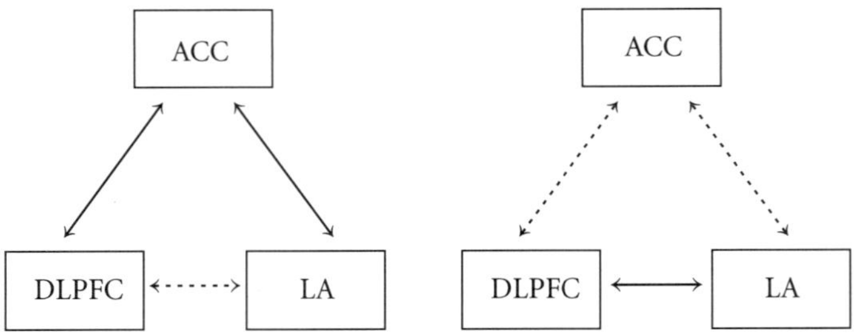

Figure 8. Neural basis of working memory for HSS and LSS.

HSS, whereas it has weakly connected with both areas. Thus, ACC appears to play a major role in the central executive for efficient attentional control under working memory task.

The working memory approach to the high-level consciousness could thus be clarified in terms of individual differences. However, further investigation is needed to resolve mystery of consciousness from the perspective of working memory.

References

Baddeley, A. (1986). *Working memory*. Oxford: Oxford University Press.

Baddeley, A. (1992). Working memory. *Science, 255*, 556–559.

Baddeley, A. (1996). Exploring the central executive. *Quarterly Journal of Experimental Psychology, 49A*, 5–28.

Braver, T. S., Barch, D. M., Gray, J. R., Molfese, D. L., & Snyder, A. (2001). Anterior cingulated cortex and response conflict: Effects of frequency, inhibition and errors. *Cerebral Cortex, 11*, 825–836.

Bunge, S. A., Klinberg, T., Jacobson, R. B., & Gabrieli, D. E. (2000). A resource model of the neural basis of executive working memory. *Proceedings of the National Academy off Sciences, USA,* 97, 3573–3578.

Bush, G., Whalen, P. J., Rosen, B. R., Jenike, M. A., Mclnerney, S. C., & Rauch, S. L. (1998). The counting stroop: An interference task specialized for functional neuroimaging: Validation study with functional MRI. *Human Brain Mapping, 6*, 270–282.

Bush, G., Luu, P., & Posner, M. I. (2000). Cognitive and emotional influences in anterior cingulated cortex. *Trends in Cognitive Science, 4*, 215–222.

Courtney, S., Ungerleider, L., Keil, K., & Haxby, J. (1996). Object and spatial visual working memory activate separate neural systems in human cortex. *Cerebral Cortex, 6*, 39–49.

Courtney, S., Petit, L., Maisog, J., Ungerleider, L., & Haxby, J. (1998). An area specialized for spatial wprking memory in human frontal cortex. *Science, 279*, 1347–1351.

Daneman, M., & Carpenter, P. (1980). Individual differences in working memory and reading. *Journal of Verbal Learning and Verbal Behavior, 19*, 450–466.

D'Esposito, M., Detre, J., Alsop, D., Shin, R., Atlas, S., & Grossman, M. (1995). The neuralbasis of the central executive system of working memory. *Nature, 378*, 279–281.

Fuster, J. (1995). *Memory in the cerebral cortex: An empirical approach to neural network in the human and nonhuman primate.* Cambridge: MIT Press.

Fuster, J. (1997). Network memory. *Trends in Neuroscience, 20*, 451–459.

Just, M., & Carpenter, P. (1992). A capacity theory of comprehension: Individual differences in working memory. *Psychological Review, 99*, 122–149.

Just, M. A., Carpenter, P. A., Keller, T. A., Eddy, W. F., & Thulborn, K. R. (1996b). Brain activation modulated by sentence comprehension. *Science, 274*, 114–116.

MacDonald III, A., Cohen, J. D., Stenger, V., & Carter, C. (2000). Dissociating the role of the dorsolateral prefrontal and anterior cingulated cortex in cognitive control. *Science, 288*, 1835–1838.

Osaka, M., & Osaka, N. (1994). Reading and working memory capacity: An examination using Reading Span Test. *Japanese Journal of Psychology, 65*, 339–345 (In Japanese).

Osaka, M., & Osaka, N. (1992). Language-independent working memory as measured by Japanese and English reading span tests. *Bulletin of the Psychonomic Society, 30*, 287–289.

Osaka, M., Nishizaki, Y., Komori, M., & Osaka, N. (2002). Effect of focus on verbal working memory: Critical role of the focus word in reading. *Memory & Cognition, 30*, 562–571.

Osaka, M., Osaka, N. Koyama, S., Okusa, T., & Kakigi, R. (1999). Individual differences in working memory and the peal alpha frequency shift on magnetoencephalography. *Cognitive Brain Research, 8*, 365–368.

Osaka, M., Osaka, N., Kondo, H., Morishita, M., Fukuyama, H., Aso, T., & Shibasaki, H. (in press). The neural basis of individual differences in working memory capacity: An fMRI study. *NeuroImage.*

Osaka, N. (1997). In the theatre of working memory of the brain. *Journal of Consciousness Studies, 4*, 332–334.

Osaka, N. (1998). Brain model of self-consciousness based on working memory. *Japanese Psychological Review, 41*, 87–95.

Osaka, N. (2000). Visual working memory. In N. Osaka (Ed.), *Brain and working memory.* Kyoto University Press (In Japanese).

Petrides, M. (1994). Frontal lobes and working memory: Evidence from investigations of the effects of cortical excisions in nonhuman primates. In F. Boller & J. Grafman (Ed.), *Handbook of neuropsychology*, Vol. 9. Amsterdam: Elsevier.

Petrides, M. (1996). Specialized systems for the processing of mnemonic information within the primate frontal cortex. *Philosophical Transactions of the Royal Society of London, B, 351*, 1455–1462.

Raichle, M. E. (1993). The scratchpad of the minds. *Nature, 363*, 583–584.

Rao, S., Rainer, G., & Miller, E. (1997). Integration of what and where in the primate prefrontal cortex. *Science, 276*, 821–824.

Rushworth, M., Nixon, P., Eacott, M., & Passingham, R. (1997). Ventral prefrontal cortex is not essential for working memory. *The Journal of Neuroscience, 17*, 4829–4838.

Smith, E. E., Geva, A., Jonides, J., Miller, A., Reuter-Lorenz, P., & Koeppe, R. A. (2001). The neural basis of task-switching in working memory: Effects of performance and aging. *Proceedings of the National Academy of Sciences, USA, 98*, 2095–2100.

Ungerleider, L. (1995). Functional brain imaging studies of cortical mechanisms for memory. *Science, 270*, 769–775.

Ungerleider, L., & Mishkin (1982). Two cortical visual systems. In D. Ingle, M. Goodale, & R. Mansfield (Eds.), *Analysis of visual behavior*. Cambridge: MIT Press.

Vogt, B. A., Finch, D. M., & Olson, C. R. (1992). Functional heterogeneity in cingulate cortex: The anterior executive and posterior evaluative regions. *Cerebral Cortex, 2*, 435–443.

Wilson, F., Oscalaidhe, S., & Goldman-Rakic, P. (1993). Dissociation of object and spatial processing domains in primate prefrontal cortex. *Science, 260*, 1955–1958.

Chapter 4

Consciousness, intelligence and creativity

A personal credo

Rodney M. J. Cotterill
Danish Technical University

Prologue

I have written this somewhat unorthodox paper in order to present my current views on consciousness, intelligence and creativity. This has become desirable because there are certain positions in some of my published works that do not have the clarity I had sought. In particular, some authors seem to believe that my theory of consciousness requires a necessarily active participation of the muscles. This would be ludicrous, and it is emphatically not the case. But my explanation of the manner in which the underlying brain circuits could support totally covert simulation of the body's muscular interaction with the environment has not always been as transparent as I had hoped. I am thus using this article to present a more coherent, yet succinct, review of my beliefs on the subject. One could call it a personal credo. And I am taking the opportunity of including a discussion of related matters. Some readers may feel outraged by the inclusion of an exhaustive personal bibliography of my work, but I could not resist the temptation of including a complete list of *all* my writings on this subject. Readers who turn to those articles will discover references to well over a thousand other authors on these subjects. It has, indeed, always been a point of honour with me to cite *all* relevant work by other people, despite the fact that this gesture has only very seldom been reciprocated.

Definition

Consciousness, a state accessible only to a creature possessing a sufficiently advanced nervous system, is intimately related to that system's sole external product: movement. It mediates acquisition and use of context-specific movement patterns, both overtly and covertly, and it thereby expands an individual creature's behavioural repertoire during its own lifetime. The efficiency of such acquisition and use is determined by the creature's intelligence, while creativity is the capacity for exploring and incorporating novel movement scenarios. Consciousness is thus a prerequisite for intelligence, and intelligence is a prerequisite for creativity.

The basic behavioural mode of all creatures

The basic behavioural mode of all creatures, even unicellular examples devoid of nervous systems, is self-paced probing of the environment. Through such movement, provoked by its internal state, the creature detects factors in the environment to which its receptors – molecules or receptor-molecule-containing nerve cells – are sensitive. Depending on how the environmental feedback influences the internal state, such detection may or may not cause a creature to react, and change its current behavioural mode. An auxiliary behavioural mode involves reaction to unprovoked external stimuli, and this is seen in most creatures, even some unicellular varieties.

The externally provoked reactions seen in some unicellular creatures are automatic and immediate. Such reactions are known as reflexes, and they are also part of the behavioural repertoire of creatures with nervous systems. When such latter creatures are too primitive to possess consciousness, the reflexes are merely inherited, though they can be modulated by environmental factors. Creatures with consciousness can expand their reactive options by augmenting their innate reflexes with more sophisticated patterns of movement, and they can do so covertly by thought, without actually performing movements overtly. Thought is thus internal simulation of a creature's interaction with the environment, and it requires special features to be present in the nervous system.

Cognition, schemata and knowledge

Because movement is a creature's only external output, cognition is best gauged through its behavioural reactions, either actual or merely potential. The unit of self-paced environmental probing is the *schema*, which is defined by its concurrently activated motor planning and anticipated environmental feedback patterns corresponding to a cognitive element. In the most primitive creatures, those patterns are embodied in cascades of pre-ordained chemical reactions, but when a nervous system is present the patterns are series of nerve impulses, schemata being stored in memory. The schema is thus not just a prescription for movement; it also implies knowledge of the movement's consequences, in terms of probable environmental feedback. Knowledge is thus inextricably linked to schemata and their activation. And knowledge is just as firmly linked to behaviour, overt or covert.

In most creatures, schemata are activated unconsciously, and the associated knowledge is implicit. But the schemata of creatures with nervous systems are differentiated, even in those not possessing consciousness. Many schemata merely embody the routine feedback caused by the creature's touching its own body surfaces, whereas the most useful schemata relate to external objects. Even though they are not conscious, therefore, such creatures are able to distinguish between self and non-self – between subject and object.

The underlying behavioural strategy for all creatures, therefore, is one of conditionally permitted movement, and the selection of a schema appropriate to the prevailing conditions, both in the environment and within the creature's own body, could be compared with the operation of a clutch. With each evolutionary advance, the clutch mechanism became more differentiated and sophisticated. When the drive provoking movement arises internally, as when hunger gives rise to foraging, the choice of schema is directed by the creature's internal state. When the drive relates to an environmental factor, the choice may consume a certain amount of time, and this is why reflexes are so useful; they provide a short cut to the appropriate reaction.

Acquisition of novel schemata

Creatures possessing that ability learn new schemata through their exploration of their environment, such exploration being either overt or covert. The individual creature thus ratchets itself upward to ever-increasingly sophisticated modes of behaviour. When a new schema is being acquired, the more elementary schemata of which it is composed are brought to the focus of attention and held in working memory. Those more elementary schemata thus become the means by which the desired end result, the new schema, is constructed. When that new schema has been incorporated into the individual's repertoire of schemata, it is available as a building block in the creature's quest for still more complicated schemata. It is in that manner that formerly explicit ends are gradually relegated to the role of implicit means, as when a child who has learned to transfer its weight from one foot to the other can employ that strategy when learning to walk.

In such an accretion process, attention and working memory could be said to conspire so as to provide the system with a central executive. But that executive would be ineffectual if it could not also draw on a mechanism for evaluating the relative merits of competing potential composite schemata. In particular, memories of the adverse outcomes of the activation of certain schemata must be made available. The executive can invoke these outcomes rapidly. The access is a top-down mechanism triggered automatically by the schemata the system is preparing to bring into play.

Raw sensations and emotions

Recalling what was stated above regarding the essential link between schemata and knowledge, we are here adding the fact that a vital aspect of that knowledge is the potential consequence of a schema's activation. This is an indispensable component of evaluation. And given that knowledge is just as inescapably linked to muscular activation, overt or covert, we see that evaluation must, perforce, implicate the body's musculature, either overtly or covertly. According to this picture, raw sensation itself is inextricably tied to schemata, and it arises from the interplay between the activation of the relevant sensory receptors and the corresponding motor sequences. The plurality of the latter should be noted. The viewing of the colour red strikes up potential links to a variety of red objects, with their differing connotations, and there is also the link to the (muscular) articulation of the word *red*. Exactly which muscular response, if any, is subsequently activated will depend upon the associated circumstances. Sensations are thus seen to be associated with the anticipation to act. Nausea, for example, is associated with the anticipation of vomiting.

The word commonly used in connection with raw sensations is qualia (singular: quale). The present author has suggested that the link between the quale and the schema is that when the activation of a schema is accompanied by attention, a quale is produced. Just as elementary schemata can be consolidated into a new schema, so can elementary qualia be envisaged as contributing to the composite quale associated with a new schema. Seen in this light, qualia add the vital ingredient of value to shemata, and they thus provide the basis of evaluation. Emotions, according to this scenario, arise from the interplay between the schemata associated with observing the creature's current situation and the schemata associated with the situation the creature's inner drives are propelling it toward. Broadly speaking, positive emotions are thus evoked when the circumstances and inner drives are in harmony, while negative emotions arise out of a conflict between circumstances and drives.

Thought and the stream of consciousness

The acquisition of new context-specific reflexes is clearly advantageous. Our familiarity with such reflexes tends to make us overlook them. We learn multiplication tables when young, for example, and use them with consummate ease throughout our lives, thereby relieving ourselves of the need to carry out simple routine calculations. We soon forget the effort we invested when first learning these aids to efficiency. And it is easy to overlook the fact that recitation of these tables involves muscular movements, namely those of the vocal apparatus.

There is great scope for conflict between the body's muscles, so their use must be under the perpetual surveillance of the above-mentioned adjudicating mechanism. The instantaneous state of the total musculature, with respect to the state of contraction and also the rate of change of contraction of each and every muscle, can be characterised by a single point moving in muscular hyperspace. Our overt behaviour is nothing more than the trajectory of this point, as time evolves. The trajectory is the outcome of a continuous series of elementary decisions, which must select the best possible muscular state under the instantaneous circumstances. The situation is made complex by the fact that the circumstances themselves can be influenced by the body's own muscular movements.

The unity of conscious experience stems from the need of the muscles to interact with the environment. At any instant, the objects and events in our surroundings are uniquely defined, and the muscular states must be similarly unique at any instant. Another way of stating this is to say that the stream of consciousness must have a unique and continuous plot. Our emotional state, at any instant, must be similarly uniquely defined, even though the subject may not be able to find words to describe it adequately. Language is often misleading in this respect. One has terms such as bittersweet, and oxymoron expressions such as terribly nice, but at the deepest level the emotional state is nevertheless unambiguously defined.

Thought, according to this theory, is simply the covert simulation of the body's interactions with the environment. It has to arise because the system needs to simulate the projected outcome of an intended movement, concurrently with the movement's actual execution. It is through detection of a potential mismatch that the above-mentioned trajectory can be prescribed, and the creature's goals achieved. The thinking process is most in evidence when the body is not simultaneously executing muscular movements, or when it is making movements not related to the current thoughts because the movements are automatic. We can act without thinking, think without acting, act and think about the act, and act while thinking about something else. This great flexibility makes considerable demands on the underlying anatomy and physiology, and it is not surprising that thought is the exclusive province of the most advanced species. In the view of the present author, thought is present in all mammals, despite the fact that they are not all imbued with the same level of intelligence. The present author does not believe that reptiles or lower species think, while he is still undecided on the issue of thought in birds.

The consciousness mechanism and creativity

Figure 1, which has been reproduced from one of the author's previous publications (Cotterill 2001b), is an overall view of a major fraction of the nervous system's anatomy. It indicates the importance of the environmental feedback that results from the movement of muscles, that feedback coming in two broad classes, namely slow and rapid. It is the latter that is believed to be vital to consciousness. And one sees from the anatomy that copies (known as efference copies) of the nerve signals dispatched to the muscles are sent to the cerebellum. It is thus not possible for the muscles to be ordered to contract without the cerebellum being informed of the fact. The mossy-fibre input to the cerebellum transmits response signals from the environment, via the various sensory receptors. And because the cerebellum captures correlations between its two types of input, that is to say between the efference copy and environmental feedback, it facilitates the gradual learning that underlies subsequent anticipation. The upshot of this is that when signals are dispatched to the muscles from the motor-directing areas, there is already an anticipation of the likely feedback. This is, of course, the gist of the schema concept.

Not all signals impinging on the sensory receptors will be the consequence of the creature's own muscular movements, of course. On the contrary, and as is indicated in Figure 1, there will also be unprovoked signals emanating from the environment. But those latter signals will not be associated with efference copy signals, so the system has a reliable means of differentiating between the signals it is receiving as a consequence of its own self-paced movements and externally originating signals. In other words, it has an automatic means of distinguishing between subject and object. It thus has a means of knowing that it is an *agent*, even though that knowledge will merely be implicit if it is not accompanied by attention.

This recalls our earlier discussion of how the creature *knows*, namely through its muscular apparatus, or a surrogate for the latter (see below). But from what has just been stated, mere knowing must not be equated with consciousness. It is rather the case that *consciousness involves knowing that one knows*. And from the above argument, one must assume that this too implicates the musculature or its cerebral representatives, namely the motor-planning areas. This harmonizes with what was stated earlier, in terms of the relationship between the stream of consciousness and the trajectory of the state point through muscular hyperspace. The perpetual corrections to that trajectory, required to achieve a match between aims and actions, are made possible by the adjudicating mechanism, which in turn relies on the existence of the monitoring mechanism mediated by the raw sensations.

As can also be seen from Figure 1, the muscle spindles contribute to the anticipatory mechanism, because they monitor the match between intended and actual muscle contraction. But the figure shows that there are cerebral routes involving loops that bypass even the spindles. There may thus be a totally covert variant of thought. But this would be able to draw only on standard schemata already acquired, unless it can invent new ones. That latter process is creativity, of course, and its underlying neuronal basis has been the subject of a lengthy analysis, based on competition between candidate schemata randomly generated (Cotterill 2001b).

A problem could arise in connection with this putative covert mechanism, because there would seem to be no obvious source of inertia. Although signals have the potential for transmitting information, they do not achieve this until they have influenced something that offers resistance, mechanical or otherwise. Radio waves, for example, are detectable only when they have impinged on an instrument in which things move, be they things as small as electrons.

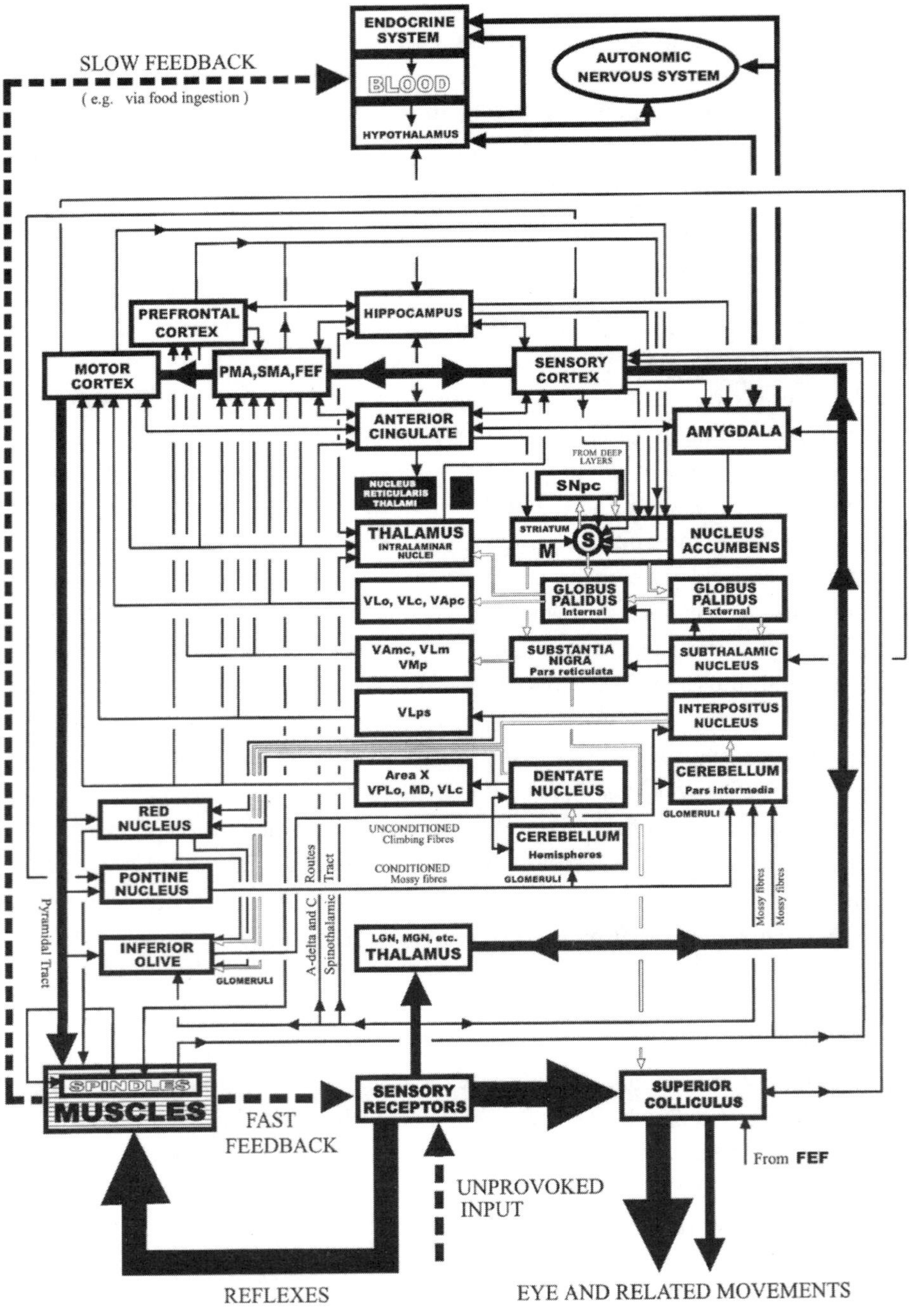
ENDOCRINE SYSTEM
BLOOD
HYPOTHALAMUS
AUTONOMIC NERVOUS SYSTEM
SLOW FEEDBACK
(e.g. via food ingestion)
PREFRONTAL CORTEX
HIPPOCAMPUS
MOTOR CORTEX
PMA,SMA,FEF
SENSORY CORTEX
ANTERIOR CINGULATE
AMYGDALA
FROM DEEP LAYERS
NUCLEUS RETICULARIS THALAMI
SNpc
THALAMUS INTRALAMINAR NUCLEI
STRIATUM
S
M
NUCLEUS ACCUMBENS
VLo, VLc, VApc
GLOBUS PALIDUS Internal
GLOBUS PALIDUS External
VAmc, VLm VMp
SUBSTANTIA NIGRA Pars reticulata
SUBTHALAMIC NUCLEUS
VLps
INTERPOSITUS NUCLEUS
Area X VPLo, MD, VLc
DENTATE NUCLEUS
CEREBELLUM Pars intermedia
GLOMERULI
RED NUCLEUS
UNCONDITIONED Climbing Fibres
CEREBELLUM Hemispheres
CONDITIONED Mossy fibres
GLOMERULI
PONTINE NUCLEUS
Pyramidal Tract
A-delta and C Routes
Spinothalamic Tract
Mossy fibres
Mossy fibres
INFERIOR OLIVE
GLOMERULI
LGN, MGN, etc. THALAMUS
SPINDLES
MUSCLES
FAST FEEDBACK
SENSORY RECEPTORS
SUPERIOR COLLICULUS
From FEF
UNPROVOKED INPUT
REFLEXES
EYE AND RELATED MOVEMENTS

What would be the analogous things in the case of purely covert simulation of the body's interaction with the environment? When the spindles are directly implicated, there is no problem of course, because the potential movement is associated with the mechanical properties of the spindles themselves. But what replaces this factor when the spindles are bypassed? I suspect that it is associated with the functioning of the premotor cortex. It is in that region of the cerebrum that candidate schemata must compete with each other, through mutual inhibition, and this process is essentially a competition for resources, the latter being ultimately governed by the movement of ions. There seems to be no reason why ions in the premotor cortex should not be placed on an equal footing with the electrons in one's radio receiver. If this view proves tenable, it would clear away any difficulty associated with purely covert simulation, and it would explain why people who are totally paralysed nevertheless appear to be capable of thought.

Dreaming

The theory described here suggests a use for the dreaming state, namely to provide a period of consolidation of novel schemata recently probed, without the potential interference of new signals from the environment. During wakefulness, the creature explores new schemata, though its deliberate or accidental putting together of series of more elementary schemata, with the associated adjudication as to suitability. As can be seen from Figure 1, the hippocampus is well situated to capture the correlations between the executed series of muscular commands and the resulting environmental feedback. And that brain component is activated during dreaming. It is thus reasonable to conclude that the hippocampus directs the conversion of candidate schemata, temporarily stored in the hippocampus itself, into permanent schemata stored in the appropriate regions of the cerebrum.

Figure 1. The connections between many of the individual components of the nervous system are shown in this schematic circuit diagram. The positioning of the various components has been idealised, in the cause of clarity, but no component has been displaced so much as to give rise to confusion. The individual lines (black for excitation and white for inhibition) indicate routes, rather than individual nerve cell axons. A full list of abbreviations is given in the author's major article published in *Progress in Neurobiology* (Cotterill 2001b).

Intelligence

If the theory summarised here turns out to be correct, it would offer a straightforward extension to intelligence. The ability of an individual to exploit the potential for novel schemata acquisition will clearly depend upon the system's efficiency. There are several contributing sub-mechanisms, such as detection of the environmental feedback provoked by the creature's self-paced muscular movements and the holding in working memory of the elementary schemata that comprise the candidate novel schema. Any lowering of efficiency of even a single sub-mechanism will impair the overall acquisition mechanism. It is the present author's view that this will be equivalent to the lowering of intelligence, it being necessary to bear in mind that this impairment may be specific regarding faculty. There seems to be no reason why such impairment could not be circumscribed, rather than global. Autism is well known to be frequently associated with at least partial depression of the intelligence level. And it has been discovered that infant autistic people display a decreased ability to concatenate elementary movements, so as to produce more complicated movement scenarios.

Such overt movements have their purely covert counterparts, of course, and it is these that must provide the basis of reason, if the views expressed in this article prove to be correct. So here too, an undiminished working memory will be essential if the power of reason is to achieve its full potential. Reason, in other words, requires the holding in working memory of competing schemata, and the capacity for juggling such schemata is clearly related to what the psychologist calls embedding. It is interesting to note, in this connection, that the digit span of autistic people is markedly lower than that measured in normal subjects.

If a creature is sufficiently intelligent, it will be aware of own ultimate demise. This is clearly the case with humans, and it might also be true of the higher primates. So consciousness – and its chief product, intelligence – could be said to be the heavy, and paradoxical, price a creature has to pay in order to enjoy its larger set of survival options.

A personal bibliography

Cotterill, R. M. J. (1989). *No Ghost in the Machine: Modern Science and the Brain, the Mind and the Soul.* London: Heinemann.

Cotterill, R. M. J. (1990). Is the brain a "lowerarchy"? In H. Bohr (Ed.), *Characterising Complex Systems*. Singapore: World Scientific.

Cotterill, R. M. J. (1994). *Autism, Intelligence and Consciousness*. Copenhagen: Munksgaard.

Cotterill, R. M. J. (1995). On the unity of conscious experience. *Journal of Consciousness Studies, 2* (4), 290–312.

Cotterill, R. M. J. (1996). Prediction and internal feedback in conscious perception. *Journal of Consciousness Studies, 3* (3), 245–266.

Cotterill, R. M. J. (1997a). On the mechanism of consciousness. *Journal of Consciousness Studies, 4* (3), 231–247.

Cotterill, R. M. J. (1997b). On the neural correlates of consciousness. *Cognitive Studies: Bulletin of the Japanese Cognitive Science Society, 4*, 31–44.

Cotterill, R. M. J. (1997c). Navigation, consciousness and the body/mind "problem". *Psyke & Logos, 18*, 337–341.

Cotterill, R. M. J. (1998). *Enchanted Looms*. Cambridge: Cambridge University Press.

Cotterill, R. M. J. (2000a). On brain and mind. *Brain and Mind, 1*, 237–244.

Cotterill, R. M. J. (2000b). Movement, acquisition of novel context-specific reflexes and amyotrophic lateral sclerosis: Reply to Jesse Prinz. *Brain and Mind, 1*, 257–263.

Cotterill, R. M. J. (2000c). Muscular hyperspace and navigation in the theatre that never closed, the cognitive bacterium, conscious unity, self-tickling, and computer simulation: Reply to Marcel Kinsbourne. *Brain and Mind, 1*, 275–282.

Cotterill, R. M. J. (2000d). Did consciousness evolve from self-paced probing of the environment, and not from reflexes? *Brain and Mind, 1*, 283–298.

Cotterill, R. M. J. (2000e). Consciousness, schemata and language. In P. Århem, C. Blomberg, & H. Liljenström (Eds.), *Disorder Versus Order in Brain Function: Essays in Theoretical Neurobiology*. Singapore: World Scientific.

Cotterill, R. M. J. (2000f). Consciousness *really* explained? *Consciousness and Cognition*, 9(2), S28–S29.

Cotterill, R. M. J. (2001a). Evolution, cognition and consciousness. *Journal of Consciousness Studies, 8* (2), 3–17.

Cotterill, R. M. J. (2001b). Cooperation of the basal ganglia, cerebellum, sensory cerebrum and hippocampus: Possible implications for cognition, consciousness, intelligence and creativity. *Progress in Neurobiology, 64*, 1–33.

Chapter 5

Cerebral physiology of conscious experience

Experimental studies in human subjects

Benjamin Libet
University of California, San Francisco

Our experimental studies attempted to discover how neural events in the brain are related to production of conscious and unconscious mental functions. Electrical stimulation of somatosensory sites with intracranial electrodes in awake human subjects, indicated that awareness required that cerebral activities last up to 500 ms. Sensory experience is thus delayed, but subjectively the timing is antedated to the earliest input to the cortex.

In a voluntary act brain activity was found to precede awareness of conscious intention by about 400 ms. Voluntary acts are thus initiated unconsciously, but conscious function could still control the outcome by veto.

The transition from an unconscious detection to sensory awareness also requires an increase in duration of cerebral activations. This "time-on" model has important implications.

Subjective experience may be viewed as an attribute of a "conscious mental field" (CMF). The CMF emerges from but is phenomenologically distinct from brain activity. The CMF could unify subjective experience and mediate the action of conscious will in neural activities. An experimental design to test this radical hypothesis is presented.

Philosophical theories and analyses of the relationship between conscious mind and neural activities in the brain have been important in examining ways of looking at this relationship (e.g. Nagel 1979). However, any theory that purports to specify how the mind and brain are actually interrelated should be testable by observations, whether experimental or descriptive. The proposal by Descartes, that the mind is located in the pineal body, was testable; unfortunately, he did not or could not test what happens upon destruction of the pineal. Our own approach has been to frame questions in terms of neuronal functions that may mediate the production of and the transition between conscious and unconscious events, and to investigate these experimentally by simultaneously observing and manipulating cerebral neuronal functions on the one hand and introspective reports of subjective experiences on the other (Libet et al. 1964; Libet 1966, 1973).

1. Experimental approach

The conscious experiences studied were either simple somatosensory ones ('raw feels') or conscious intentions/wishes to initiate a simple voluntary action (sudden flexion of a wrist). These psychologically simple events minimized potential complications from emotional or other impacts on the validity of introspective reports, they were amenable to experimental tests of their reliability, and areas of cerebral cortex involved in their mediation were available for electrophysiological study with intracranial and extracranial electrodes in awake human subjects.

In our experimental approach to the issue I set for myself two epistemological principles which I believe must be followed, for a workable, operationally valid criterion of a conscious experience:

Principle 1. Introspective report of the experience

The stubborn fact is that conscious experience or awareness is directly accessible only to the subject having that experience, and not to an external observer. Consequently, only an introspective report by the subject can have primary validity as an operational measure of a subjective experience. The 'transmission' of an experience by a subject to an external observer can, of course, involve distortions and inaccuracies. Indeed there can be no absolute certainty (objectively demonstrable by externally observable events) about the validity of the report (Libet 1987). Of course, absolute certainty does not hold even for physical observations. But in all social interactions we accept introspective reports as evidence of a subject's experience, though we may subject the report to some analyses and tests which affect our acceptance of their validity. This same general approach can be applied more rigorously in a scientific investigation. The meaning of what is communicated depends in part upon the degree to which both individuals have had similar or related experiences. A congenitally completely blind person can never share the conscious experience of a visual image, regardless of how detailed a verbal description he or she is given by a sighted individual. The same limitation applies to all experiences in less dramatic, more subtle ways. For example, electrical stimulation of the somatosensory cortex can produce sensations related to but sufficiently different from those generated by normal sensory input, so that the subjects could only relate some roughly understandable approximation of these experiences to the experimental observer, in whom similar modes of sensory generation had never been employed (Libet et al. 1964; Libet 1973).

An important corollary of this principle (introspective reports) is that any behavioral evidence which does not require a convincing introspective report cannot be assumed to be an indicator of conscious, subjective experience. This is so regardless of the purposeful nature of the action or of the complexity of cognitive and abstract problem-solving processes involved, since all of these can and often do proceed unconsciously, without awareness by the subject. Similarly, studies of signal detection should not be confused with those of conscious experience. The forced choice responses in signal detection studies can be made independent of introspective awareness of the signal, although they may be excellent indicators of whether some type of detection has occurred. There is, in fact, evidence that signal detection can occur with signals that are distinctly below the activations required for any conscious awareness of the signal (see Libet et al. 1991). Indeed, most sensory signals probably do not reach conscious awareness; but many of them lead to modified responses and behaviors, as in the tactile and proprioceptive signals that influence simple everyday postural and walking activities and which have therefore clearly been detected and utilized in complex brain functions.

The requirement of a reportable introspective experience also implies that it is difficult for nonhuman animals, even monkeys or chimpanzees, to supply convincingly valid responses for our purposes. Such animals may in fact have conscious experiences, but the question is – how can these be studied? Complicated cognitive and purposeful behaviors can proceed even in human beings without introspective awareness of them (e.g. Shevrin & Dickman 1980; Velmans 1991; Libet 1973), and so we cannot safely assume that such behaviors in animals are expressing subjective experience. Some investigators and writers have proposed that adaptive, problem-solving behaviors in animals indicate conscious thought and experience. But even in humans, the most complex problem solving, even of mathematical problems, can and often does proceed at unconscious levels, as has been repeatedly described by many creative thinkers, artists and others.

Principle 2. No a priori rules for mind-brain relationship

Mental events and externally observable (physical) events constitute two categorically separate kinds of phenomena. They are mutually irreducible categories in the sense that one of them cannot, *a priori*, be described in terms of the other. There are no *a priori* rules or laws that can describe the relationship between neural brain events and subjective mental events. No doubt rules do exist, but they must be discovered. They can only be established by simultaneous observation of both categories of event. It also follows that even a complete knowledge of all the neural physical events and conditions (in the unlikely case that were possible) would not, in itself, produce a description of any associated subjective experience (Libet 1966; Nagel 1979). This would constitute a flat rejection of a reductionist view (e.g. Dennett 1991) that an adequate knowledge of neuronal functions and structure would be sufficient for defining and explaining consciousness and mental activities.

2. The neural time factor in perception

Cerebral sensory stimulation

We embarked on our long range investigation by attempting to determine what kinds of activation patterns, in a primary sensory area of cerebral cortex, seemed uniquely required for the production of a conscious sensory experience. This was done by controlling the electrical stimuli applied to somatosensory cortex in awake human subjects, via electrodes implanted subdurally during therapeutic neurosurgical procedures. It was already well known that electrical stimulation there can elicit somatic sensations which are subjectively referred by the patient to a localized body structure on the side contralateral to the stimulus site. In adopting this experimental approach we were in accord with a suggestion by the great British neuroscientist Lord Adrian (1952). He stated "I think there is reasonable hope that we may sort out the particular activities that coincide with quite simple mental processes like seeing or hearing" (or, in our studies, bodily sensations). "At all events that is the first thing a physiologist must do if he is trying to find out what happens when we think and how the mind is influenced by what goes on in the brain".

It became evident that duration (of the train of repetitive brief electrical pulses) was a critical factor in producing a conscious sensory response to stimulation of sensory cortex (see Libet 1973, review). A surprisingly long stimulus

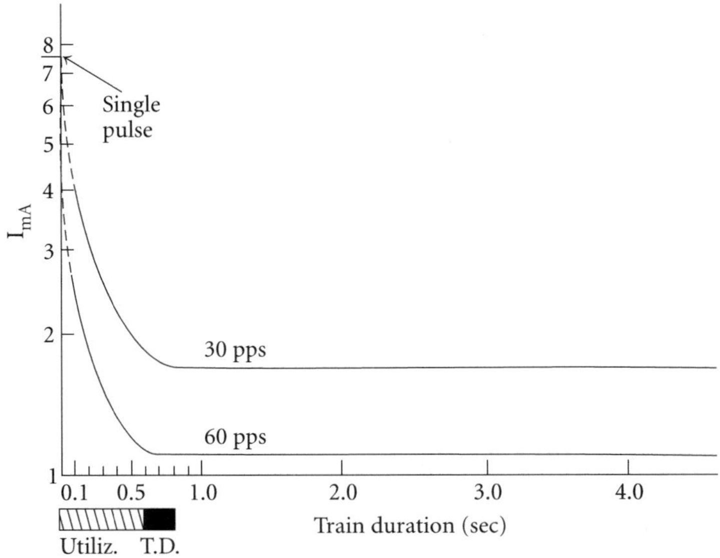

Figure 1. Temporal requirement for stimulation of the somatosensory (SI) cortex in human subjects. The curves are schematic representations of measurements in five different individuals, which were not sufficient in each case to produce a full curve individually. Each point on a curve would indicate the combination of intensity (I) and train duration (TD) for repetitive pulses that is just adequate to elicit a threshold conscious sensory experience. Separate curves shown for stimulation at 30 pulses per second (pps) and 60 pps. At the liminal I (below which no sensation was elicited even with long TDs), a similar minimum TD of about 600 ms ±100 ms was required for either pulse frequency. Such values for this "utilization TD" have since been confirmed in many ambulatory subjects with electrodes chronically implanted over SI cortex and in the ventrobasal thalamus (from Libet 1973).

duration of about 500 ms was required when the peak intensity of the pulses was at the liminal level for producing any conscious sensation (see Figure 1). Changes in frequency of the pulses, electrode contact area or polarity, and so forth did not affect the requirement of a long duration. Nor was this requirement unique to locating the stimulus on the surface of cerebral cortex. The same temporal relationship was found with stimulating electrodes placed at any point in the cerebral portion of the direct ascending sensory pathway that leads up to the somatosensory cortex, that is, in subcortical white matter, or in the sensory relay nucleus in ventrobasal thalamus, or in the medial lemniscus bundle that ascends from nuclei in the medulla to the thalamic nucleus. However, the requirement does not hold for the spinal cord below the medulla, or indeed for peripheral sensory nerves from the skin itself; there, a single pulse can be effective for sensation even at near liminal intensity (Libet et al. 1967).

The average 500 ms requirement in cerebral structures is not a rigidly fixed one; with intensities stronger than the liminal level, briefer stimulus durations can be effective. But a single pulse could not elicit a sensation even with very strong intensities (up 40 times the liminal intensity needed for a 500 ms stimulus train). It appears probable that a minimum duration of about 100 ms is required for even the strongest inputs.

Must a single pulse stimulus to the skin also induce prolonged neural responses of cerebral cortex in order to elicit a conscious sensation?

The answer to this had to be more indirect, but at least three lines of evidence have been convincingly affirmative on it. These will be only listed here without all the experimental details.

i. Skin stimuli that were too weak to evoke the appropriate later components of event-related potentials at the cerebral cortex but could still evoke a primary response did not elicit any conscious sensation (Libet et al. 1967). Pharmacological agents (e.g. atropine or general anesthetics) that depress these late components also depress or abolish conscious sensory responses.
ii. The sensation induced by a single stimulus pulse to the skin can be retroactively enhanced by a stimulus train applied to somatosensory cortex, even when the cortical stimulus *begins* 400 ms or more after the skin pulse (Libet 1978; Libet et al. 1992). This indicates that the content of a sensory experience can be altered while the experience is 'developing' during a roughly 500 ms period before it appears.
iii. Reaction times to a peripheral stimulus were found to jump discontinuously, from about 250 ms up to more than 600–700 ms, when subjects were asked deliberately to lengthen their reaction time by the smallest possible amount (Jensen 1979). This surprising result can be explained by assuming one must first become aware of the stimulus signal in order to delay one's response deliberately; if up to 500 ms is required to develop that awareness, then the reaction time cannot be deliberately increased by lesser amounts.

Subjective referral of the timing vs. the neural delay, for a sensory experience

A neural delay of up to about 500 ms would mean that we do not experience the sensory world immediately, in real time. But, intuitively we do not experience sensory events as having delays relative to their actual occurrence. If there is a substantial neural delay required before achieving a sensory experience or awareness, is there actually a corresponding delay in the subjective timing of that experience?

In a direct test of this issue we paired a skin stimulus (single pulse) with a cortical stimulus train of pulses (at liminal intensity, 500 ms duration required), and asked the subject to report which of these stimulus-induced sensations came first. Subjects in fact reported that the skin-induced sensation came first even when the skin pulse was applied a few hundred ms after the onset of the cortical train (Figure 2). That is, there appeared to be no appreciable subjective delay for the skin-induced sensation relative to the delayed cortically induced sensation. The clue to solving this paradox lay in the differences between electrophysiological responses to primary somatosensory SI cortical stimuli and those to skin stimuli. Cortical stimuli did not elicit any primary evoked potentials at SI; but each skin pulse does produce this component (10-20 ms delay), as well as later components out to 500 ms or more (Libet et al. 1975).

This led us to propose that (1) the early primary evoked neural response acts as a timing signal, and (2) there is a subjective referral of the timing of the skin-induced experience, from its actually delayed appearance back to the time of the initial fast primary evoked response of the cortex (which has only a 10–20 ms delay) (Figure 3). The experience would thus be subjectively "antedated" and would seem to the subject to have occurred without any delay. No such antedating would occur with the cortical stimulus since the normal initial or primary response, to a sensory volley ascending from below, is not generated by our surface-cortical stimuli. This hypothesis was tested and confirmed as follows: A skin stimulus was paired with one in the direct subcortical pathway that leads to sensory cortex (medial lemniscus), Figure 2 (Libet et al. 1979). Unlike the cortical stimulus, the subcortical one in medial lemniscus does elicit the initial primary response in the cortex, the putative timing signal, with each pulse; but, unlike the skin stimulus, it resembles the cortical one in its requirement of a long (up to 500 ms) train duration of pulses. As predicted by the hypothesis, the subcortically induced sensation was reported to have no delay relative to the skin-induced sensation, even though it was empirically es-

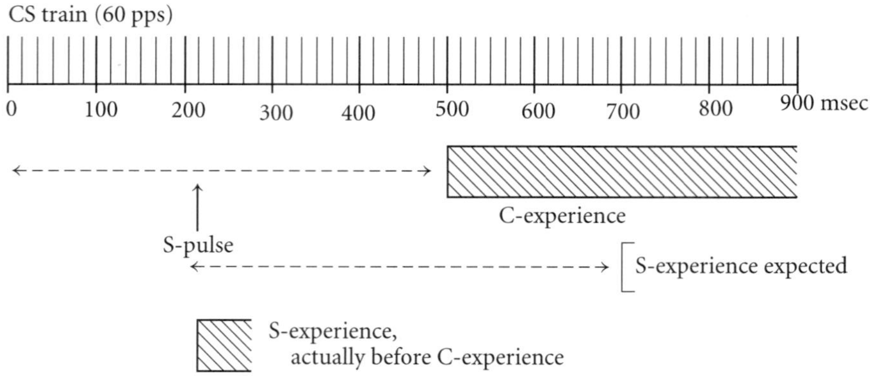

Figure 2. Diagram of an experiment on the subjective time order of two sensory experiences, one elicited by a stimulus train to the SI cortex (labelled C) and the other by a threshold pulse to skin (S). CS consisted of repetitive pulses (at 60/sec) applied to the postcentral gyrus (C), at the lowest (liminal) peak current sufficient to elicit any reportable conscious sensory experience. The sensory experience for CS ('C-experience") would not be initiated before the end of the utilization-train duration (average about 500 ms), but then proceeds without change in its weak subjective intensity for the remainder of the applied liminal CS train (see Libet et al. 1967; Libet 1966, 1973). The S-pulse, at just above threshold strength for eliciting conscious sensory experience, is here shown delivered when the initial 200 ms of the CS train have elapsed (in other experiments, it was applied at other relative times, earlier and later). If S were followed by a roughly similar delay of 500 ms of cortical activity before "neuronal adequacy" is achieved, initiation of S-experience might have also been expected to be delayed until 700 ms of CS had elapsed. In fact, S-experience was reported to appear subjectively before C-experience.
For the test of the subjective antedating hypothesis the stimulus train was applied to the medial lemniscus (LM) instead of to somatosensory cortex (CS in the figure). The sensation elicited by the LM stimuli was reported to begin before that from S, in this sequence; this occured in spite of the empirically demonstrated requirement that the stimulus train in LM must persist for 500 ms here in order to elicit any sensation (see text). (From Libet et al. 1979, by permission of *Brain*.)

tablished that the subcortically induced experience could not have appeared before the end of the stimulus train (whether 200 or 500 ms, depending on intensity); see Figure 2.

Such subjective referral thus serves to correct the temporal distortion of the real sensory event, a distortion imposed by the cerebral requirements of a neural delay for the experience. An analogous subjective correction occurs for the spatial distortion of the real sensory image, imposed by the spatial representa-

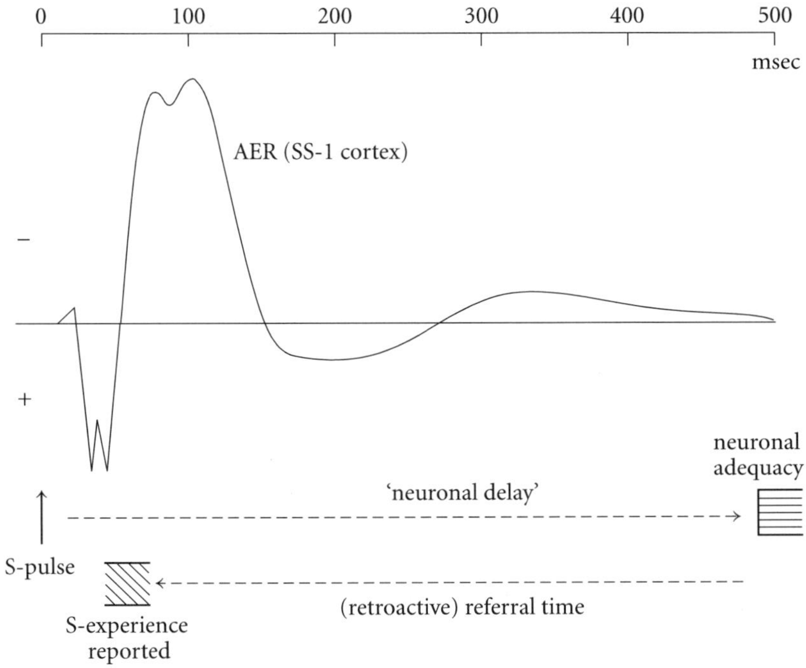

Figure 3. Diagram of hypothesis for subjective referral of sensory experience backward in time. The average evoked response (AER) based on 256 averaged responses delivered at 1/sec, recorded at the somatosensory cortex (labelled SS-1 here), was evoked by pulses just suprathreshold for sensation and delivered to the skin of the contralateral hand. Below the AER, the first line shows the approximate delay in achieving the stage of neuronal adequacy that appears (on the basis of other evidence) to be necessary for eliciting the sensory experience. The lower line shows the postulated retroactive referral of the subjective timing of the experience, from the time of neuronal adequcy backwared to some time associated with the primary surface-positive component of the evoked potential. The primary component of the AER is relatively highly localized to an area on the contralteral postcentral gyrus in these awake human subjects. The secondary or later components, especially those following the surface-negative component after the initial 100 to 150 ms of the AER, are more widely distributed over the cortex and more variable in form, even when recorded subdurally (see, for example, Libet et al. 1972). This diagram is not meant to indicate that the state of neuronal adequacy for eliciting conscious sensation is restricted to neurons in the primary SI cortex of postcentral gyrus; on the other hand, the localized primary component or "timing signal" for retroactive referral of the sensory experience is more closely restriced to this SI cortical area. (From Libet et al. 1979, by permission of *Brain*.)

tion of the image in the responding neurons of the sensory cortex. Nowhere in the brain is there a response configuration that matches the sensory image as perceived subjectively. Subjective referrals of the timing of the spatial configuration of an experience are clear examples of an event in the mental sphere that was not evident in or predictable by knowledge of the associated neural events.

3. Initiation of a voluntary act

Does an endogenous experience, one arising in the brain without any external instigators, also require a substantial duration of neural activity before the awareness appears? The possibility to investigate this issue had its roots in the discovery by Kornhuber and Deecke (1965) and Deecke, Grotzinger, and Kornhuber (1976) that an electrical change (the "readiness-potential" or RP) is recordable at the vertex starting up to a second or more prior to a simple "self-paced" movement. Accepting this electrical change as an indicator of brain activity that is involved in the onset of a volitional act, we asked the question: Does the conscious wish or intention to perform that act precede or coincide with the onset of the preparatory brain processes; or does the conscious intention follow the cerebral onset?

We first established that an RP is recordable even in a fully spontaneous voluntary act, with an average onset of about –550 ms, before the first indication of muscle action ("O time"). (The presence of a component of "preplanning" when to act makes RP onset even earlier; this probably accounts for the longer values given by others.) We then devised and tested a method whereby clock-time indicators of the subject's first awareness of his/her wish to move (W) could be obtained reliably. RPs (neural processes) and Ws (the times of conscious intention) were then measured simultaneously in large numbers of these simple voluntary acts (Libet et al. 1983a).

The results clearly showed that onset of RP precedes W by about 350 ms in freely spontaneous voluntary acts (Figure 4). This means that the brain has begun the specific preparatory processes for the voluntary act well before the subject is even aware of any wish or intention to act; that is, the volitional process must have been initiated unconsciously (or nonconsciously). This sequence is also in accord with the principle of a substantial neural delay in the production of a conscious experience generally.

Is there any role for the conscious function in voluntary action? It is important to note that conscious intention to act (W) does appear about 150–200 ms before the act. Thus, although *initiation* of the voluntary act is an uncon-

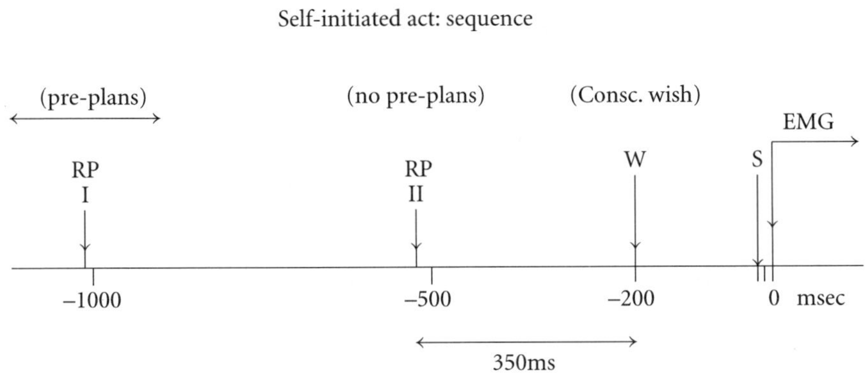

Figure 4. Diagram of sequence of events, cerebral and subjective, that precede a fully self-initiated voluntary act. Relative to 0 time, detected in the electromyogram (EMG) of the suddenly activated muscle, the readiness potential (RP) (an indicator of related cerebral neuronal activities) begins first, at about –1050 ms when some preplanning is reported (RP I) or at about –550 ms with spontaneous acts lacking immediate preplanning (RP II). Subjective awareness of the wish to move (W) appears at about –200 ms, some 350 ms after onset even of RP II but well before the act (EMG). Subjective timings reported for awareness of the randomly delivered S (skin) stimulus average about –50 ms relative to actual delivery time. (From Libet 1989, by permission of Cambridge University Press.)

scious function of the brain, the conscious function still appears in time to potentially affect the *outcome* of that volitional process. That is, the conscious function could potentially either promote the culmination of the process into action, or it could potentially block or veto the final progression to the motor act (Libet 1985). Free will is thus *not* excluded by our findings.

4. How does the brain distinguish unconscious from conscious mental functions?

Many, if not most, mental functions or events proceed without any reportable awareness, i.e. unconsciously or non-consciously. These apparently include cognitive detection of sensory signals and appropriate behavioral responses to them, for example in the blindsight phenomenon (Weiskrantz 1986) and in word primings (see Holender 1986), and in the cerebral initiation of a voluntary act (Libet 1985). There is descriptive evidence for unconscious processing of even complex functions, as in problem-solving or intuitive and creative thinking. (Recall the anecdotes by the great French mathematician Poincare about his unconscious solving of major problems, the solution of which then sprang fully formed into his consciousness.) This broad view of unconscious mental functions extends far beyond a more limited one of Freudian "repression" of emotionally difficult thoughts. However, it would not include those cerebral operations which are "non-mental", in the sense that they do not ever achieve a potentiality to rise into awareness; these would include control processes for blood pressure, heart rate, automatic postural adjustments, etc. On the other hand, the simplest kinds of mental functions can be accompanied by awareness/subjective experience, like the awareness of a localized tap on the skin or of a few photons of light on the retina, etc. It is not, then, simply the complexity or creativeness of a mental function that imparts to it the quality of subjective awareness of what is going on. The cerebral code for the distinction between the appearance or absence of awareness in any mental operation would seem to require a mediating neuronal mechanism uniquely related to awareness *per se*, rather than to complexity, etc. In view of our finding of a substantial temporal requirement for the production of even a threshold sensory awareness, I have proposed a 'time-on' theory that provides one potentially controlling factor for the neural distinction between conscious and unconscious mental events.

Cerebral "time-on" theory

Our evidence indicated that a minimum activity time ("time-on") is required to produce a conscious experience, whether the experience is exogenous (sensory) or endogenous (conscious intention to act voluntarily). We had also found that activations with durations less than that minimum time did in fact elicit considerable neuronal responses. These were evident in the DCR responses recorded from sensory cortex during a train of stimulus pulses nearby (Libet 1973), and in the evoked cortical potentials elicited by a single pulse in sensory thalamus that produced no awareness (Libet et al. 1967). It seemed attractive to hypothesize that neuronal activation times to produce an unconscious mental function could be less than those for a conscious one. The transition from an unconscious to a conscious function could then be controlled simply by an adequately increased duration of the appropriate neuronal activity. I have called this proposal the "time-on" theory.

A direct experimental test confirmed the theory, at least for the case of somatosensory inputs (Libet et al. 1991). Stimulus trains, 72 pps at just above liminal intensity, were delivered to ventrobasal thalamus, with train durations that varied in different trials, from 0 to 760 ms. With a forced choice test of the subject's ability to detect the presence of a stimulus, correct responses in 50% of trials would be expected on chance alone. Statistical analysis of the thousands of trials in nine subjects showed that correct detection (well above 50%) occurred even when subjects were completely unaware of any sensation and were guessing. The mean train duration for all trials in which there was correct-detection-with-*no*-awareness was compared to the mean train duration for all trials in which there was correct-detection *and* awareness (of "something", that is, at the uncertain level of awareness). To achieve even this uncertain awareness when being correct required an additional stimulus duration of 375 ms. These results confirmed the "time-on" theory.

The results demonstrate that *detection* of a signal can be sharply distinguished from *awareness* of the signal. This distinction is often overlooked in cognitive studies. Strictly speaking, all cognitive experiments in nonhuman animals are studying detection, as no introspective reports or other credible criteria of awareness are available. Since detection with no reportable awareness is clearly possible even in human subjects, one cannot regard detection of a signal by a monkey as valid evidence for the additional feature of awareness of that signal. The results also argue against a view that conscious experience arises when a sufficient level of neuronal complexity is achieved. The degree of complexity in the detections with no awareness was presumably at least as great as that for detections with awareness; the significant distinction was the duration of the appropriate neural activities, not their complexity.

5. How would the minimum duration of neuronal activity lead to a conscious experience?

There appear to be at least three general options: (1) The repetition of appropriate neuronal activities for up to 0.5 s is finally integrated to elicit some *specific neuronal* event that 'signifies' or is accompanied by a conscious event. (2) The substantial minimum *duration* of neuronal repetition could itself constitute the 'code' for the appearance of an accompanying conscious event. (3) The required duration of activities may be necessary to form at least a *short-term memory* of the event; without such a memory there would be nothing reportable after the process is over.

First option

The possibility that the generation of a conscious event is mediated by an integrative mechanism, sensitive simply to intensity and duration of the neuronal activities and leading to a specific neuronal event,does not agree with available evidence: (i) When stimuli to the ventrobasal thalamus or to somatosensory cortex (postcentral gyrus) were just below the liminal intensity for eliciting sensory experience, then no conscious sensation was elicited even by stimulus durations of 5 s or longer (Libet et al. 1964; Libet 1973). Such 'subliminal' intensities are not below threshold for eliciting neuronal responses; substantial electrophysiological responses of large populations of neurons are recordable with each such 'subliminal' stimulus pulse. Were simple integration of intensity and duration the controlling mechanism, a sufficiently long train duration of stimulus pulses would be expected to become effective for awareness. (ii) At a liminal intensity which becomes effective with an average 0.5 s of train duration, the neuronal responses recordable electrically at the cortex exhibit no progressive alteration during the train and no unique event at the end of the 0.5 s train (Libet 1973, 1982). Obviously, not all the possible neuronal activities were recordable, but this evidence offers no support for a progressive integrative factor. (iii) The minimum train duration that can elicit awareness, when the intensity is raised as high as possible, has not been firmly established, although it would appear to be in the order of 100 ms. However, it was empirically shown that a single stimulus pulse localized to the medial lemniscus could not elicit any conscious sensation no matter how strong (Libet et al. 1967); this was true even when the intensity of the single pulse was 20–40 times the strength of the liminal I (liminal I is the minimum peak current to elicit sensation when delivering a train of pulses with duration >0.5 s). The effectiveness of 10 pulses (at 20/s) at liminal I contrasted with the ineffectiveness of a single pulse at 40 times this liminal I, argues against a simple integrative mechanism.

Second option

Mildly supportive of the second option is the fact that no specific or unique neuronal event has thus far been found in recordings of 'direct cortical responses' to stimulation of the cortex (Libet 1973) or in event-related potentials (e.g. Libet et al. 1975). Admittedly, many possibilities of an undetected neuronal event remain. The decision between the two options remains open, pending further experimental investigations.

Third option

It is agreed that formation of short-term memory is a necessary ingredient in order for the minimum activity time to produce a reportable experience. It has been argued that awareness of a stimulus may appear almost immediately but that this conscious experience would be forgotten and become not reportable without the formation of a memory by the required 0.5 s of activity (Dennett 1991; Dennett 1993). This would contrast with our proposal that awareness appears only after the 0.5 s delay. The following evidence makes (a) the "immediate awareness plus memory" hypothesis untenable and supports our proposal (b) of a delay for awareness:

i. The reported subjective timing of the sensation elicited by a stimulus at the surface of SI cortex in fact showed a delay approximately equal to the 0.5 s duration of a train of liminal stimulus pulses. That contrasted with the subjective timing of awareness of a skin pulse; that was reported to appear almost immediately after the single liminal pulse to the skin. But the subjective immediacy of the skin sensation was shown to be due to a *subjective antedating* of the experience that actually did not appear until after a 0.5 s of later neuronal responses by the SI cortex (Libet et al. 1979; Libet 1993a). The immediacy of the subjective timing of the skin pulse is a function of the early primary evoked potential; without the latter the subjective timing even of a single skin pulse is delayed by the same 0.5 s seen with the stimulus at SI cortex. Memory formation is indeed necessary to make any sensation reportable, but it is the presence of the primary evoked cortical response that accounts for the subjective immediacy for the actually delayed sensory awareness, not memory formation. This was shown in a striking way with stimuli in medial lemniscus (Libet et al. 1979). Stimulus trains here produce a primary cortical response with *each* pulse. The resulting sensation was reported to appear immediately at the outset of the train, even though 200 or 500 ms of train duration was empirically required to elicit any sensation!
ii. Secondly, although development of a very short-term memory of an experience is of course required for reportability some seconds after the stimulus, the process for producing that memory need not reside in the specific neural activity that is actually involved in producing the experience. Our demonstration (Libet et al. 1991) that very brief thalamic stimuli can be detected without generating awareness provides evidence against proposal (a). Because our subjects made their forced-choice responses after the same post-stimulus period of some seconds, whether or not they were

aware of the stimulus, the same short-term memory of the signal was obviously produced, even by cortical stimuli of 100 ms or less. Therefore a short-term memory was essential in all trials, whether there was or was not awareness of the stimulus. That is, the production of a short-term memory trace did not distinguish between an unconscious and a conscious response to the stimulus, even though durations of effective inputs to SI cortex were greatly different for the two cases. It might be argued that there are two different modes of memory formation involved, so-called implicit memory of unconscious events and explicit memory for a reportable conscious event (Roediger 1990; Squire et al. 1993). Some evidence indicates that the two forms of memory formation may utilize different neural pathways. If such a distinction applies to our unconscious detections vs. conscious awarenesses of a stimulus, it would mean that formation of an implicit memory required a much smaller duration of the inputs to SI cortex than does formation of explicit memories. One would thus have to add the unlikely ad hoc hypothesis that intrinsic memory formation is a function of the much shorter durations of cortical activations that are adequate for an unconscious detection. Even then, that would not explain the immediate subjective timing coupled with a required 0.5 s stimulus duration, as found with stimuli in medial lemniscus.

iii. Proposal (a), that invokes a forgetting of an immediate awareness during the initial few hundred msec of cerebral activations after a skin pulse, is not experimentally falsifiable. As Velmans pointed out (1993: 145–146), "when establishing a stimulus threshold one gradually turns up the intensity until the subject reports feeling the stimulus. It is standard procedure to assume the subject's report is accurate", etc. But according to proposal (a), the "inability to report the stimulus might have resulted from rapid forgetting". Proponents of (a), like Dennett (1993) "could extend that claim to any reports that subjects make about not having experienced something. So, in spite of any claims subjects make to contrary, Dennett could maintain his position (that there was an experience but it was forgotten). That makes Dennett's position "unfalsifiable"". When a proposal is unfalsifiable it is worthless as a scientific statement; it is not better than any other unfalsifiable statement or hypothesis, including metaphysical speculations and beliefs.

6. Some implications of "time-on" theory

Cerebral representation

If the transition from an unconscious to a conscious mental function could be dependent simply on a suitable increase in duration of certain neural activities, then both kinds of mental functions could be represented by activity in the same cerebral areas. Such a view would be in accord with the fact that the constituents and processes involved in both functions are basically similar, except for the awareness quality.

All-or-nothing character of awareness

If the transition to and production of awareness of a mental event occur relatively sharply, at the time a minimum duration of neuronal activities is achieved, this suggests that an awareness appears in an all-or-nothing manner (Libet 1966). That is, awareness of an event would not appear weakly at the onset of an appropriate series of neural activities and gradually develop to full awareness. Conscious experience of events, whether initiated exogenously or endogenously, would thus have a unitary discontinous quality. This would be opposed to the continuous "stream of conscious" nature postulated by William James and assumed in many present theories of the nature of consciousness. It is, however, in accord with a postulate of unitary nature for mental events adopted by Eccles (1990) as part of his theory for mind-brain interaction.

Filter function

It is generally accepted that most sensory inputs do not achieve conscious awareness, even though they may lead to meaningful cerebral responses and can, in suitable circumstances (of attention, and so forth) successfully elicit conscious sensation. The "time-on" requirement could provide the basis for screening inputs from awareness, if the only inputs that elicit awareness are those that induce the minimum duration of appropriate activities. Such a requirement could prevent conscious awareness from becoming cluttered and permit awareness to be focused on one or a few event or issues at a time.

Delayed experience versus quick behavioral responses

Meaningful behavioral responses to a sensory signal, requiring cognitive and conative processing, can be made within as little as 100–200 ms. Such responses have been measured quantitatively in reaction time tests and are apparent in many kinds of anecdotal observations, from everyday occurrences (as in driving an automobile) to activities in sports (as when a baseball batsman must hit a ball coming at him in a tortuous path at 90 miles per hour). If actual conscious experience of the signal is neurally delayed by several hundred milliseconds, it follows that these quick behavioral responses are performed unconsciously, with no awareness of the precipitating signal, and that one may (or may not) become conscious of the signal only after the action. Direct experimental support of this was obtained by Taylor and McCloskey (1990), who showed that the reaction time for a visual signal was the same whether the subject reported awareness of the signal or was completely unaware of it (owing to the use of a delayed masking stimulus).

Subjective timing of neurally delayed experience

Although the experience or awareness of an event appears only after a substantial delay, there would ordinarily be a subjective antedating of its timing back to the initial fast response of the cortex, as discussed above (Libet et al. 1979). For example, a competitive runner may start within 100 ms of the starting gun firing, before he is consciously aware of the shot, but would later report have heard the shot before starting.

There is another facet to this issue: For a group of different stimuli, applied synchronously but differing in location, intensity and modality, there will almost certainly be varying neural delays at the cortex in the times at which these different experiences can appear. This asynchrony could lead to a subjective temporal jitter for the group of sensations that were initiated synchronously. However, if each of these synchronously appearing experiences is subjectively antedated to its initial fast cortical response, they would be subjectively timed as being synchronous, without subjective jitter; the differences among their initial fast cortical responses are approximately 10 ms and too small for subjective separation in time.

Unconscious mental operations proceed speedily

If there is virtually no minimum "time-on" requirement for unconscious (or nonconscious) mental processes in general, then these could proceed quickly, in contrast to conscious events. This feature is obviously advantageous, not only for fast meaningful reactions to sensory signals but also for the more general complex, intuitive, and creative mental processes, many of which are deemed to proceed unconsciously. Conscious evaluation would be expected, according to the theory, to be much slower.

Opportunity for modulation of a conscious experience

It is well known that the content of the introspectively reportable experience of an event may be modified considerably in relation to the content of the actual signal, whether this be an emotionally laden sensory image or endogenous mental event (which may even be fully repressed, in Freud's terms). For a modulating action by the brain to affect the eventual reportable experience, some delay between the initiating event and the appearance of the conscious experience of it seems essential. The "time-on" theory provides a basis for the appropriate delays. We have some direct experimental evidence for such modulatory actions on the awareness of a simple sensory signal from the skin: An appropriate cortical stimulus begun 400 ms or more after the skin pulse could either inhibit or enhance the sensory experience (Libet et al. 1992; Libet 1978, 1982).

Finally, it is important to recognize that the neural time factors in conscious experience could not have been discovered and established without the experimental and falsifiable tests that pitted actual neural activations against the accompanying subjective reports with human subjects. They could not have been discovered by constructing theories based on the previous knowledge of brain processes.

7. "Time-on" theory, conscious control and free will

The experimental evidence indicates that a voluntary act is initiated in the brain unconsciously, before the appearance of the conscious intention. The question then arises, what role, if any, does the conscious process itself have in volitional actions? (In this, we are considering only the processes immediately involved in the performance of a voluntary movement. The issue of conscious planning of how, whether and when to act is a separate one.) Clearly, free will or free choice of whether 'to act now' could not be the initiating agent, contrary to one widely held view.

We must distinguish the initiation of a process leading to a voluntary action from control of the outcome of that process. The experimental results showed that a conscious wish to act appeared at about –200 ms, i.e. before the motor act, even though it followed the onset of the cerebral process (readiness potential) by about 350 ms (see Figure 4). This provides a period during which the conscious function could potentially determine whether the volitional process will go on to completion. That could come about by a conscious choice either to promote the culmination of the process in action (whether passively or by a conscious 'trigger'), or to prevent the progress to action by a conscious blockade or veto. The potential for such conscious veto power, within the last 100–200 ms before an anticipated action, was experimentally demonstrated by us (Libet et al. 1983b). It is also in accord with common subjective experiences, that one can veto or stop oneself from performing an act after a conscious urge to perform it has appeared (even when the urge appears suddenly and spontaneously).

Even if we assume that one can extrapolate these results to volitional acts generally, they do not exclude a possible role for free will. However, the potential role of free will would be constrained; free will could no longer be an initiator of the voluntary act, but only a controller of the outcome of the volitional process, after the individual becomes aware of an intention or wish to act. In a general sense, free will could only select from among the brain activities that are a part of a given individual's constitution.

If we generalize the 'time-on' theory for producing awareness, a serious potential difficulty arises if the theory should also apply to the initiation of the *conscious control* of a volitional outcome. If the conscious control function itself were to be initiated by unconscious cerebral processes, one might argue there is no role at all for conscious free will, even as a controlling agent. However, conscious control of an event is not the same as becoming aware of the volitional intent. Control implies the imposing of a change, in this case after the appearance of the conscious awareness of the wish to act. In this sense, conscious control may not necessarily require the same neural 'time-on' feature that may precede the appearance of awareness *per se*. There is presently no specific experimental test of the possibility that conscious control requires a specific unconscious cerebral process to produce it. Given the difference between a control and an awareness phenomenon, an absence of the requirement for conscious control would not be in conflict with a general 'time-on' theory for awareness. Thus, a potential role for conscious free will would remain viable in the conscious control, though not in the initiation, of a voluntary act.

8. Other philosophical issues for possible experimental study

One of the most mysterious and seemingly intractable problems in the mind-brain relationship is that of the *unitary and integrated nature of conscious experience*. This phenomenon is somehow a product of a brain with an estimated 200 billion neurons, each of which may have thousands of interconnections with other neurons. There is increasing evidence that many functions of the cerebral cortex are localized, apparently organized into specialized columns of neurons. In spite of the enormously complex array of structures and functions, whatever does reach awareness is experienced as unified and integrated. A second apparently intractable question is issue B – *can the conscious mental process actually influence or control some activities of the cerebral neurons?*

Both of these issues have been the subjects of many philosophical analyses and proposals since Descartes and earlier, but this is not the place to review all that. Rather, our approach as scientific investigators should be to inquire whether experimentally testable hypotheses can be generated in attempting to deal with these two issues. On issue A, Eccles (Popper & Eccles 1977: 362) has proposed, based on a dualist-interactionist view, that 'the experienced unity comes not from a neurophysiological synthesis but from the proposed integrating character of the self-conscious mind'. This view had, in princple, been expressed by Sherrington (1940), and also by Sperry (1980) and Doty (1984). For Sperry and Doty, however, the mind doing the integrating was viewed in monistic terms, as an emergent property of brain processes. On issue B, both Eccles and Sperry have proposed the the 'mental sphere' could influence neuronal function, but in dualistic vs. monistic terms, respectively. The view held probably by most neuroscientists (and perhaps modern philosophers) is a monist-determinsitic one, in which conscious mental experience is simply an 'inner aspect' of brain function (identity theory); it is fully determined by knowable physicochemical processes in the brain, and its apparent ability to influence brain function is a subjective illusion with no actual causal powers. All of these views are in fact philosophical theories. Although each has explanatory power and each can be shown to be compatible with (not falsified by) the available evidence, none has been subjected to appropriate or adequate experimental testing in a format that could potentially falsify it.

A recent neurophysiological observation may offer one possible route to investigating issue A, i.e. experiential integration or, as it is often referred to, the 'binding problem'. This is the discovery of widespread synchronization of oscillatory neuronal electrical potentials in response to a visual image (Gray & Singer 1989; Singer 1990). It has led to some speculation that a 'correlation' model (based on the widely occuring synchronization of some activities) might represent the neural coding for a unified mental image in an otherwise chaotic background. This speculation is still to be tested. I would, however, note that any experimental test should be careful to distinguish between 'binding' at a cognitive level (that may or may not involve conscious experience) and the binding that refers to unity experienced in awareness. A meaningful response to a visual image that contains forms, colours, shades of intensity, etc. requires an integration of these properties for either kind of binding, with or without awareness. A test for cognitive binding using a behavioural response in a monkey may not distinguish conscious from unconscious mental functions.

The Conscious Mental Field (CMF)

As one possible experimentally testable solution to both features of the mind-brain relationship, I would propose that we may view conscious subjective experience as if it were a field, produced by appropriate though multifarious neuronal activities of the brain (Libet 1993b, 1994).

A chief quality or attribute of the conscious mental field (CMF) would be that of a unified or unitary subjective experience. A second attribute would be a causal ability to affect or alter neuronal function. The additional meaning or explanatory power of describing subjective experience in terms of a CMF will become more evident with the proposed experimental testing of the theory. That is, the CMF is proposed as more than just another term for referring to 'unified subjective experience'.

The putative CMF would *not* be in any category of known physical fields, such as electromagnetic, gravitational, etc. The conscious mental field would be in a phenomenologically independent category; it is not describalbe in terms of any externally observable physical events or of any known physical theory as presently constituted. In the same sense as for all subjective events, the CMF would be detectable only in terms of the subjective experience, accessible only to the individual who has the experience. An external observer could only gain valid direct evidence about the conscious mental field from an introspective report by the individual subject. In this respect the conscious mental field would differ from all known physical fields, whose existence and characteristics are derived from physical observations.

On the other hand, the proposed CMF should be viewed as an operational phenomenon, i.e. as a working and testable feature of brain function. It is not proposed as a view of the metaphysical origin and nature of the mind; indeed, it could be shown to be potentially compatible with virtually any philosophical mind-brain theory. The CMF may be viewed as somewhat analogous to known physical force fields (Libet 1997). For example, a magnetic field is produced by electric current flowing in a conductor, but it can in turn influence the flow of the current. However, as indicated, the CMF cannot be observed directly by external physical means.

Experimental design to test CMF theory

The theory of a CMF makes crucial predictions that can, at least in principle, be tested experimentally. If local areas of cerebral cortex could independently contribute to or alter the larger, unitary CMF, it should be possible to demonstrate such contributions when (a) that cortical area is completely isolated or cut off from neuronal communication with the rest of the brain, but (b) the area remains *in situ*, alive and kept functioning in some suitable manner that sufficiently resembles its normal behaviour. The experimental prediction to be tested would be as follows: Suitable electrical and/or chemical activation of the isolated tissue should produce or affect a conscious experience, even though the tissue has no neural connections to the rest of the brain. Possibilities of spread of influences from the isolated block via physical non-neural paths (e.g. electric current flow) would have to be controlled for. If a subjective experience is induced and reportable within a second or so, that would tend to exclude spread by chemical diffusion or by changes in vascular circulation or in contents of circulating blood as a cause (see Ingvar 1955b).

Suitable neuronal isolation could be achieved either (a) by surgically cutting all connections to the rest of the brain, but leaving sufficient vascular connects and circulation intact, or (b) by temporarily blocking all nerve conduction into and out of an area.

A slab of cerebral cortex can be neurally isolated surgically, remaining in place but viable by retaining its blood supply as the only connection with the rest of the brain. This is accompllished by making all of the cuts subpially. Studies of the electrophysiological activity of such isolated cortex *in situ* have been reported (Kristiansen & Courtois 1949; Burns 1951, 1954; Echlin et al. 1952; Ingvar 1955a, 1955b; Goldring et al. 1961). The basic method involved introducing a narrow curved blade through an opening in an avascular area of the pia-arachnoid membrane. This could undercut a block or slab of cortex and, by bringing its tip up to meet the pia at some distance away, also cut the connections to adjacent cortex. Isolation of cortical slab has also been performed in human subjects, by Echlin et al. (1952), with both general and local anaesthesia (patient awake). They reported an immediate reduction but not complete abolition of rhythmic electrical activity (EEG) in the area. After 20 min., paraxysmal bursts of high voltage activity appeared. This kind of seizure pattern in normal brain is usually associated with disruption or distortion of normal functions and, in the motor area, convulsive motor actions. There was no spread of activity from the isolated slab to surrounding areas.

The physiological properties of the isolated slab are obviously immediately altered because of the sudden loss of all inputs. For example, it is well known that destruction of the reticular activation system in the brain stem results in a coma; this afferent input would have to be properly excited so as to 'wake up' the isolated slab of cortex. Some procedures to restore some levels of sufficiently normal activity would be necessary. These could involve local electrical stimulation (e.g. Libet et al. 1964) or the application of exciting chemical agents. Chemical stimulation of isolated cortex has already been studied (Kristiansen & Courtois 1949; Echlin et al. 1952; Rech & Domino 1960). With longer term chronic isolation, the nerve fibre inputs and their synaptic contacts with cells in the slab would degenerate and no longer provide these normal structural contacts. The studies proposed in this paper would be better carried out in the acute phase, during the initial period after isolation. Indeed, with the afferent cut axons still viable and potentially functional, they could be utilized to restore some degree of neural inputs by electrically stimulating them within the slab in a highly localized and controlled fashion.

A test of the causal ability of the putative CMF to affect neuronal functions is already implicit in the test just described for the existence of the CMF. If stimulation of the isolated cortical slab can elicit an introspective report by the subject, that could only come about if the CMF could activate the appropriate cerebral areas required to produce the verbal report.

General conclusions on CMF theory

Suppose that the experimental results prove to be positive, i.e. suitable stimulation of the neurally isolated cortex elicits some reportable subjective response that is not attributable to stimulation of adjacent non-isolated cortex or of other cerebral structures. That would mean that activation of a cortical area can contribute to overall unified conscious experience by some mode other than by neural messages delivered via nerve conduction etc. This would provide crucial support of the proposed field theory, in which a cortical area can contribute to or affect the larger conscious field. It would provide an experimental basis for a unified field of subjective experience and for mental intervention in neuronal functions.

With such a finding one may ask, what would be the role for all the massive and complex neural interconnections, cortico-cortical, cortical-subcortical and hemisphere to hemisphere? An answer might be – to subserve all the cerebral functions other than that directly related to the appearance of the conscious subjective experience and its role in conscious will. It should be noted that all cognitive functions (receipt, analysis, recognition of signals etc.), information storage, learning and memory, processes of arousal and attention and of states of affect and mood, etc. are *not* proposed as functions to be organized or mediated by the postulated CMF (conscious mental field). In short, it is only the phenomenon of conscious subjective experience, associated with all the complex cerebral functions, that is modelled in the CMF, in an admittedly speculative manner.

It may be easy to dismiss the prospect of obtaining 'positive' results in the proposed experimental tests, since such results would be completely unexpected from prevalent views of brain functions based on physical connectivities and interactions. But the improbability of positive results is strictly a function of existing views which do not deal successfully with the problems of unity of subjective experience and of apparent mental controls of brain processes. The potential implications of the CMF theory and of the positive results it predicts are clearly profound in nature. On those grounds, and because the proposed experiments are in principle workable although difficult, the proposed experimental design should merit a serious place in investigations of the mind-brain problem.

References

Adrian, E. D. (1952). What happens when we think. In P. Laslett (Ed.), *The Physical Basis of Mind*. Oxford: Blackwell.

Burns, B. D. (1951). Some properties of isolated cerebral cortex in the unanesthetized cat. *Journal of Physiology (Lond.)*, *112*, 156–175.

Burns, B. D. (1954). The production of after-bursts in isolated unanesthetized cerebral cortex. *Journal of Physiology (Lond.)*, *125*, 427–446.

Deecke, L., Grotzinger, B., & Kornhuber H. H. (1976). Voluntary finger movement in man, cerebral potentials and theory. *Biological Cybernetics, 23*, 99–119.

Dennett, D. (1991). *Consciousness Explained.* Boston: Little, Brown & Co.

Dennett, D. (1993). Discussion of Libet, B. (1993). The neural time factor in conscious and unconscious events in Ciba Foundation Symposium #174 *Experimental & theoretical studies of consciousness* (pp. 123–146). Chechester: Wiley.

Doty, R. W. (1984). Some thoughts and some experiments on memory. In L. R. Squire & N. Butters (Eds.), *Neuropsychology of Memory* (pp. 330-339). New York: Guilford.

Eccles, J. C. (1990). A unitary hypothesis of mind-brain interaction in the cerebral cortex. *Proceedings of the Royal Society of London Ser B Biological Sciences, 240*, 433–451.

Echlin, F. A., Arnett V., & Zoll, J. (1952). Paroxysmal high voltage discharges from isolated and partially isolated human and animal cerebral cortex. *Electroencephalography & Clinical Neurophysiology, 4*, 147–164.

Goldring, S., OLeary, J. L., Homes, T. G., & Jerva, M. J. (1961). Direct response of isolated cerebral cortex of cat. *Journal of Neurophysiology, 24*, 633–650.

Gray, C. M., & Singer, W. (1989). Stimulus-specific neuronal oscillations in orientation columns of cat visual cortex. *Proceedings of the National Academy of Sciences USA, 86*, 1698–1702.

Holender, D. (1986). Semantic activation without conscious identification in dichotic listening, parafoveal vision, and visual masking: A survey and appraisal. *Behavioral & Brain Sciences, 9*, 1–66.

Ingvar, D. (1955a). Electrical activity of isolated cortex in the unanesthetized cat with intact brain stem. *Acta Physiologica Scandinavica, 33*, 151–168.

Ingvar, D. (1955b). Extraneuronal influences upon the electrical activity of isolated cortex following stimulation of the reticular activating system. *Acta Physiologica Scandinavica, 33*, 169–193.

Kornhuber, H. H., & Deecke, L. (1965). Hirnpotentialanderungen bei Willkurbewegungen und passiven Bewegungen des Menschen: Bereitschaftpotential und reafferente Potentiale. *Pfluegers Arch Gesamte Physiol Menschen Tiere, 284*, 1–17.

Kristiansen, K., & Courtois, G. (1949). Rhythmic electrical activity from isolated cerebral cortex. *Electroencephalography & Clinical Neurophysiology, 1*, 265–272.

Libet, B. (1966). Brain stimulation and the threshold of conscious experience. In J. C. Eccles (Ed.), *Brain and Conscious Experience* (pp. 165–181). Berlin: Springer-Verlag.

Libet, B. (1973). Electrical stimulation of cortex in human subjects, and conscious sensory aspects. In A. Iggo (Ed.), *Handbook of Sensory Physiology* Vol. 2 (pp. 743–790). New York: Springer-Verlag.

Libet, B. (1978). Neuronal vs. subjective timing for a conscious sensory experience In P. A. Buser & A. Rougeul-Buser (Eds.), *Cerebral Correlates of Conscious Experience* (pp. 69–82). Amsterdam: Elsevier Science Publishers.

Libet, B. (1982). Brain stimulation in the study of neuronal functions for conscious sensory experiences. *Human Neurobiology, 1*, 235–242.

Libet, B. (1985). Unconscious cerebral initiative and the role of conscious will in voluntary action. *Behavioral & Brain Sciences, 8*, 529–566.

Libet, B. (1987). Consciousness: Conscious, subjective experience. In G. Adelman (Ed.), *Encyclopedia of Neuroscience.* Boston: Birkhauser.

Libet, B. (1993a). The neural time factor in conscious and unconscious events. In Ciba Foundation Symposium #174 *Experimental & Theoretical Studies of Consciousness* (pp. 123–146). Chichester: Wiley.

Libet, B. (1993b). *Neurophysiology of Consciousness.* Boston: Birkhäuser.

Libet, B. (1997). Conscious mind as a force field. *Journal Theoretical* Biology, *185*, 137–138.

Libet, B., Alberts, W. W., Wright, E. W., Delattre, L. D., Levin, G., & Feinstein, B. (1964). Production of threshold levels of conscious sensation by electrical stimulation of human somatosensory cortex. *Journal of Neurophysiology, 27*, 546–578.

Libet, B., Alberts, W. W., Wright, E. W., & Feinstein, B. (1967). Responses of human somatosensory cortex to stimuli below threshold for conscious sensation. *Science, 158*, 1597–1600.

Libet, B., Alberts, W. W., Wright, E. W., Lewis, M., & Feinstein, B. (1975). Cortical representation of evoked potentials relative to conscious sensory responses and of somatosensory qualities in man. In H. H. Kornhuber (Ed.), *The Somatosensory System.* Stuttgart: Thieme Verlag.

Libet, B., Wright, E. W. Jr., Feinstein, B., & Pearl, D. K. (1979). Subjective referral of the timing for a conscious sensory experience: A functional role for the somatosensory specific projection system in man. *Brain, 102*, 191–222.

Libet, B., Gleason, C. A., Wright, E. W., & Pearl, D. K. (1983a). Time of conscious intention to act in relation to onset of cerebral activities (readiness-potential); The unconscious initiation of a freely voluntary act. *Brain, 106*, 623–642.

Libet, B. Wright, E. W., Jr., Gleason, C. A. (1983b). Preparation- or intention-to-act, in relation to pre-event potentials recorded at the vertex. *Electroencephalography & Clinical Neurophysiology, 56*, 367–372.

Libet, B., Pearl, D. K., Morledge, D. M., Gleason, C. A., Hosobuchi, Y., & Barbaro, N. M. (1991). Control of the transition from sensory detection to sensory awareness in man by the duration of a thalamic stimulus. The cerebral time-on factor. *Brain, 114*, 1731–1757.

Libet, B., Wright, E. W. Jr., Feinstein, B., & Pearl, D. K. (1992). Retroactive enhancement of a skin sensation by a delayed cortical stimulus in man: Evidence for delay of a conscious experience. *Consciousness and Cognition, 1*, 367–375.

Nagel, T. (1979). *Mortal Questions.* Cambridge: Cambridge University Press.

Popper, K. R., & Eccles, J. C. (1977). *The Self and its Brain.* Heidelberg: Springer.

Rech, R. H., & Domino, E. F. (1960). Effects of various drugs on activity of the neuronally isolated cerebral cortex. *Experimental Neurology, 2*, 364–378.

Roediger, H. L. (1990). Implicit memory: Retention without remembering. *American Psychologist, 45*, 1043–1056.

Sherrington, C. S. (1940). *Man on his Nature.* Cambridge: Cambridge University Press.

Shevrin, H., & Dickman, S. (1980). The psychological unconscious: A necessary assumption for all psychological theory? *American Psychologist, 35*, 421–434.

Singer, W. (1991). Response synchronization of cortical neurons: An epiphenomenon or a solution to the binding problem. *IBRO News, 19*, 6–7.

Sperry, R. W., Gazzaniga, M. S., & Bogen, J. E. (1969). Interhemispheric relationships: The neocortical commissures: Syndromes of hemisphere disconnection. In P. J. Vinken & G. W. Bruyn (Eds.), *Handbook of Clinical Neurology,* Vol. 4 (pp. 273–290). Amsterdam: North Holland.

Squire, L. R., Knowlton, B., & Musen, G. (1993). The structure and organization of memory. *Annual Review of Psychology, 44*, 453–495.

Taylor, J. L., & McCloskey, D. I. (1990). Triggering of preprogrammed movements as reactions to masked stimuli. *Journal of Neurophysiology, 63*, 439–446.

Velmans, M. (1991). Is human information processing conscious? *Behavioral & Brain Sciences, 14*, 651–669.

Velmans, M. (1993). Discussion in Ciba Foundation Symposium #174 *Experimental and Theoretical Studies of Consciousness* (pp. 123–146). Chichester: Wiley.

Weiskrantz, L. (1986). *Blindsight: A case study and implications*. Oxford: Clarendon Press.

Part II

Neuronal psychophysics

Chapter 6

Neural mechanisms of perceptual organization

Nikos K. Logothetis, David A. Leopold, and David L. Sheinberg
Max-Planck Institute for Biological Cybernetics, Tübingen

1. Sensation and perception

Research over many years showed that perception is not simply determined by the patterns of neural activity in the eye's retina. The brain allows experience and expectation to play an important role in organizing sensory information, so that we do not "see" the data, but rather we use them to draw inferences as to what lies before us. Figure 1 illustrates this point clearly. The lines in Figure 1a might at first appear as a set of nonsense lines, just as the patches in Figure 1b might seem like splotches of meaningless ink. However, prior experience with these figures, or the mention of "a woman and a wash bucket", or a "dog in the yard", can dramatically alter one's perception of the very same pictures. Similarly, Figure 1c can be seen as an either an Indian or as an Eskimo with his back turned to the viewer, Figure 1d has two possible perspectives, and Figure 1e – the celebrated "faces-and-vase" figure-ground reversal introduced by the Danish psychologist Edgar Rubin in 1915 – can be perceived as either a goblet or a pair of faces. The perceptual changes occurring when viewing ambiguous figures have been shown to occur even when these figures are stabilized in the retina and thus even under conditions that the retinal stimulus remains entirely unchanged. Finally, in Figure 1f, the absence of visual stimulation appears to serve as input to our perception, as we see a white "filled" circle of high contrast in the middle and a concentric ring around it. Clearly our brain sees more than our eyes. The latter receives patterns of energy that are converted into patterns of neural excitation, while the former seldom sees such patterns. Instead the brain sees objects. Of obvious importance is the question: How does the brain

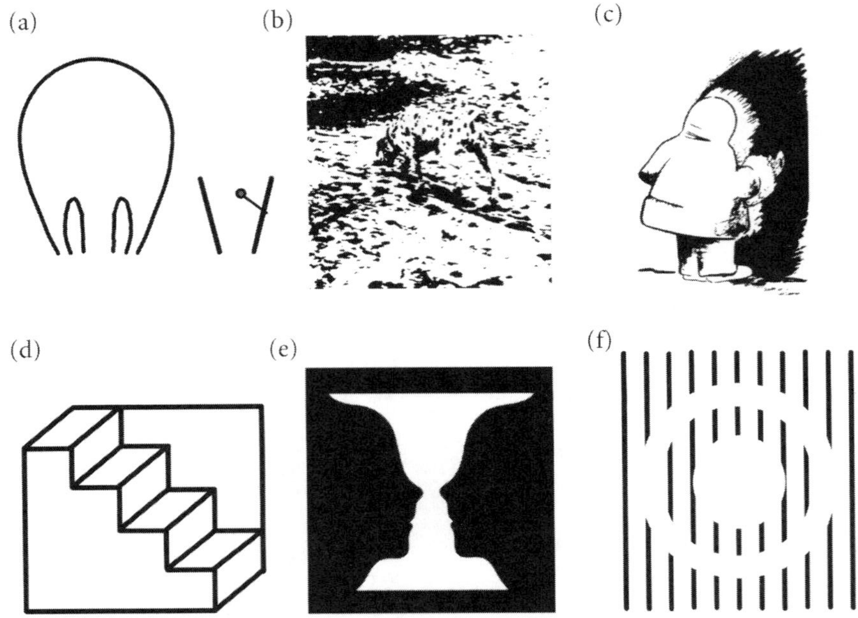

Figure 1. Stimuli used for the study of perception.

derive descriptions of definite objects out of the hodgepodge of visual elements provided through the sensory organs?

Research, initiated by the pioneering work of David Hubel and Torsten Wiesel, has shown that the visual cortex has all the machinery requisite for the formation of neural descriptions of objects. Neurons are topographically organized, show high selectivity for distinct stimulus attributes, and possess a receptive field complexity that increases in successively higher visual cortical areas. Cells in the retina and the dorsal lateral geniculate nucleus – a small thalamic structure that receives direct input from the retina and relays it to the primary visual cortex – respond to light spots; neurons in primary visual cortex respond selectively to the orientation or motion direction of line-segments; and cells in the inferior temporal cortex – a large area in the temporal lobe – may respond selectively to very complex patterns, including animate objects such as faces or body parts.

Loosely speaking, retinal processing is optimized for the detection of intensity and wavelength contrast, while early cortical areas extract fundamental stimulus descriptors, such as orientation and curvature, spatial frequencies and stimulus velocities. Higher visual areas in both the parietal and temporal lobes process information about the spatial relationships and the identity of visual objects. Yet, despite the plethora of data on the properties of individual neurons, we know very little about how their responses contribute to unified percepts, and how these percepts lead to semantic knowledge of objects. Not only do we not know how the inputs of a "face-selective" neuron in the inferior temporal cortex are organized, and thus how such a neuron comes to acquire such an amazing configurational selectivity, but we do not even know how the responses of such selective cells relate to perceptual experience. Do such cells mediate the recognition of an object? Do they represent the stage at which information is routed to the motor planning or motor areas? It is a notable fact that many of these neurons respond vigorously when presented a specific stimulus even when the animal is anesthetized. Do they then really relate to conscious perception? Which cells code for this "hodgepodge" of visual primitives and which relate directly to our knowledge of familiar objects?

Visual stimuli such as the ambiguous figures shown in figure 1 are excellent tools for addressing such questions. Normally, when we look steadily at a picture of a real object, the information received by the retina remains largely constant, as does the perception of the object, presumably because the richness of information derived by integrating a large number of visual cues establishes an unambiguous, single interpretation of a scene. In such cases neural responses that may underlie the perception of a stimulus are confounded with the sensory responses to the stimulus or parts thereof. When the visual cues provided, however, do not suffice for a single interpretation, then rival possibilities can be entertained and perception becomes ambiguous, swiftly switching between two or more alternatives without concomitant changes in the message received from the eye. Classical examples of figures eliciting different perceptions are the figure-ground and depth reversals shown in Figures 1d and 1e. The question of interest is: Would a cell, which responds selectively to, say, the profile of a face, discharge action potentials only when the faces in figure 1e are perceived, despite the fact that the pattern that can be interpreted as face is always available to the visual system? Addressing directly this question in invasive, laboratory animal experiments is extremely difficult for two reasons. First, the subject, presumably a monkey, must learn to report subtle configurational changes for one of the handful of known multistable stimuli. Second, individual neurons must be isolated that specifically respond to this stimulus, with the hope that alternate perceptual configurations differentially activate the cell.

Fortunately, perceptual bistability can also be elicited by simply presenting a conflict to the two eyes, where the monocular images differ substantially in their spatial organization, color, or direction of motion. Rarely if ever will nonmatching stimuli be binocularly fused into a stable coherent stimulus. Instead each monocular pattern takes its turn at perceptual dominance, only to be overtaken by its competitor after a number of seconds. This phenomenon, known as binocular rivalry, was first noted over two centuries ago (DuTour 1760) and its phenomenology has been studied extensively over the past three decades in the field of binocular vision (for a review see Blake 1989). Psychophysical experiments initially suggested a peripheral inhibitory mechanism for rivalry, specifically involving competition between the two monocular pathways. Rivalry has therefore been generally considered a form of interocular, rather than perceptual, competition. In other words, perception of a stimulus was thought to amount to "dominance" of the eye viewing this stimulus. Yet, we have recently shown that the notion of "eye dominance" fails to account for the long periods of perceptual dominance of a stimulus presented alternately to one eye and then to the other (Logothetis et al. 1996).

Binocular rivalry: Interocular or interstimulus competition?

Human observers were presented incongruent stimuli via a mirror stereoscope – a device permitting dichoptic stimulation of each eye under normal vergence conditions – and were asked to report periods of sustained dominance of a left or right tilted grating pattern by pressing and holding the left and right computer-mouse buttons respectively. Our experimental paradigm is outlined in Figure 2. As in many other experiments on binocular rivalry, the visual stimuli consisted of square patches of sinusoidal gratings that were orthogonally oriented in the two eyes. However, in contrast to all other experiments, the two monocular stimuli in our paradigm were exchanged between eyes every third of a second, resulting in periodic orientation reversals of each monocular view. In addition, the gratings were flickered on and off at a frequency of 18 times in a second in order to minimize the perception of transients caused by the physical exchanges. What would one expect to see under these stimulation conditions? Traditional wisdom would predict that perception will be dominated by a grating regularly switching orientations, as would be seen if the subject closed one eye. In contrast, if the perceptual rivalry were some form of competition between the central representations of the stimuli, one would expect slow alternations in perceived orientation that are uncorrelated to the physical exchange of the monocular stimuli. In our experiments, observers reported indeed seeing prolonged periods of unitary dominance of each orientation, lasting up to several seconds.

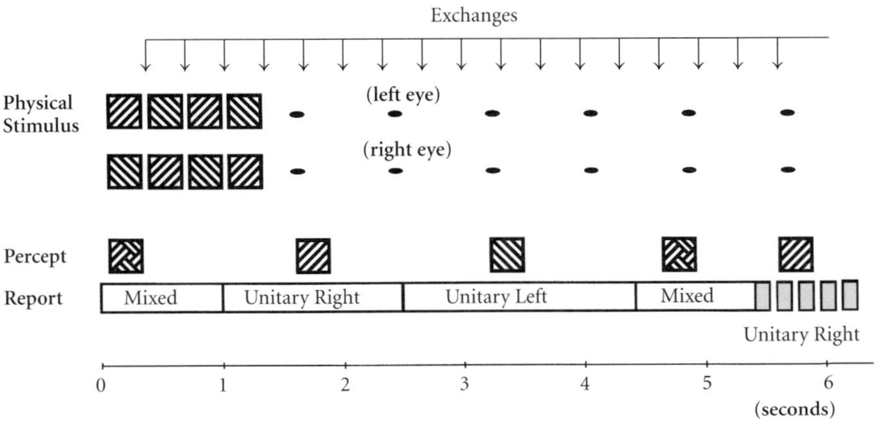

Figure 2. Psychophysical "switch" paradigm. The stimulus consisted of a pair of sinusoidal gratings of 20% contrast and 2.5 cycles per degree spatial frequency, and which were orthogonally oriented in the two eyes. The stimuli were flickered on and off at 18 Hz, and exchanged between the eyes each 333 ms. Despite the constantly reversing stimulus, perception was dominated by prolonged periods of leftward and rightward unitary dominance, with intervening periods of mixed rivalry. Subjects viewed the stimulus for two minute periods and held down buttons on a computer mouse to indicate leftward, rightward or mixed perceptual dominance.

The mean dominance duration exceeded two seconds, spanning seven physical exchanges of the gratings. Subjects reported that they rarely if ever saw a grating rapidly changing its orientation at the exchange frequency, as "eye dominance" would predict, and which could easily be observed by closing one eye. In order to compare rivalry during the "switch" paradigm with conventional rivalry, the statistics of the alternation phases were evaluated. First, successive dominance phases were shown to be independent both by autocorrelation analysis and the Lathrop test for sequential dependence (Lathrop 1966) in agreement with conventional rivalry (Fox & Herrmann 1967). Second, the distribution of these durations (Figure 3c, pooled for 3 subjects) was found to closely resemble those obtained during conventional rivalry for a human (Figure 3a) and a monkey (Figure 3b), both of which match those previously reported in the literature (Fox & Herrmann 1967; Leopold et al. 1995; Leopold & Logothetis 1995; Leopold & Logothetis 1996; Logothetis & Leopold 1995; Walker 1975).

Finally, the effects of the strength of one stimulus on the mean dominance and suppression of each were examined. The contrast of one of the orientations

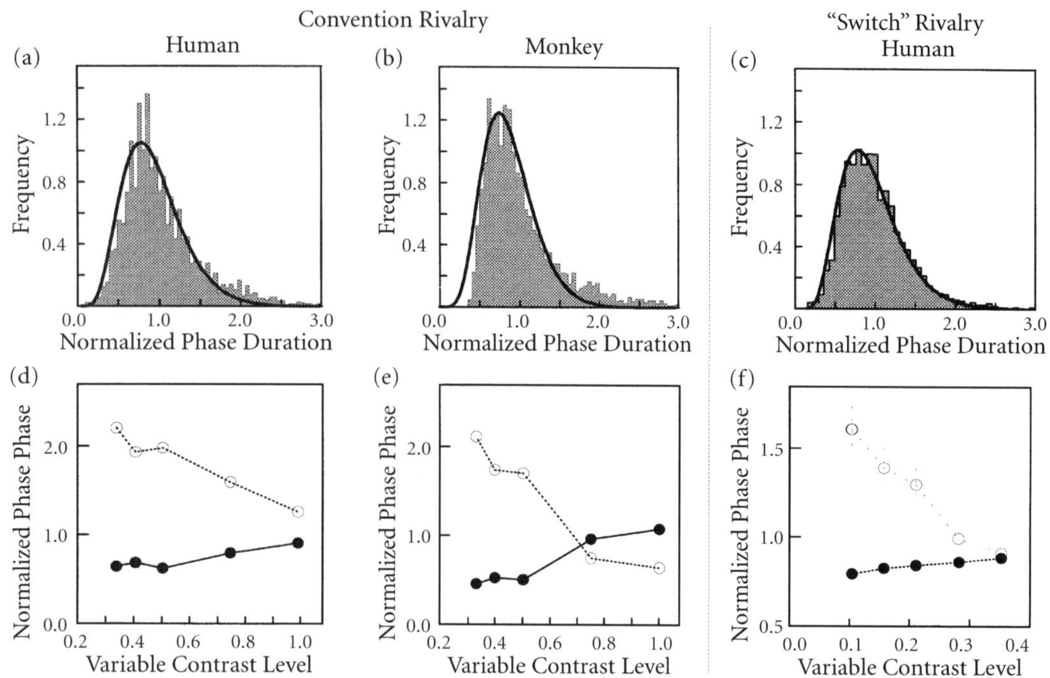

Figure 3. Statistics of perceptual rivalry alternations. The temporal dynamics of rivalry during the switch paradigm were identical with conventional rivalry. (a–c) Distribution of dominance phase durations expressed as a fraction of their mean. Data is shown for conventional rivalry from a human (a) and monkey (b), and for switch rivalry pooled from a three humans (c). The thin dark lines illustrate the approximation of the histogram data with a gamma function. (d–f) Effect of contrast of each rivalling stimulus. The abscissa shows the contrast of one stimulus, while the other was held fixed. Open circles represent the mean dominance for the fixed contrast stimulus, and closed circles for the variable stimulus. This is shown for conventional rivalry from a human (d) and monkey (e) and for switch rivalry averaged from three humans (f). The fixed contrast values were held at 1.0 and 0.35 for the conventional and switch rivalry, respectively.

was systematically varied while the other was held constant. Again, the pattern during "switch" rivalry (Figure 3f, average of 3 subjects) resembles those from a human (Figure 3d) and a monkey (Figure 3e) during conventional rivalry. In all cases the contrast of one orientation primarily affects the mean dominance time of the other orientation. Note that in the "switch" condition each eye sequentially sees both the fixed- and variable-contrast stimuli, and it is therefore the strength of the competing stimuli rather than the strength in the two eyes that governs this effect. The results of this analysis show that the rivalry experienced when monocular stimuli are continually swapped between the eyes is indistinguishable from conventional binocular rivalry.

Our finding clearly suggests that the perceptual alternations experienced during binocular rivalry involve perturbations of the same neural machinery involved in other multistable perceptual phenomena, such as monocular rivalry (Campbell & Howell 1972) and ambiguous figures, which incidently show similar dynamics to binocular rivalry (Borsellino et al. 1972). Thus dichoptic stimulation, in which arbitrary combinations of conflicting stimuli can be brought into competition for dominance, can be an excellent tool for the physiological study of perceptual organization and visual awareness in experimental animals.

2. Cell responses in early visual cortex to ambiguous stimuli

To examine the neural responses in primary visual cortex and the early extrastriate visual areas (of visual areas surrounding and receiving input from primary visual cortex) we trained monkeys to report the perceived orientation of a stimulus under congruent and dichoptic stimulation conditions (Leopold & Logothetis 1996). During the initial shaping and training phase each monkey was shown monocular grating patterns on a computer monitor and taught to press one of two levers according to whether the tilt of the grating was left or right of vertical. The animal eventually learned to respond to several successive orientation changes, receiving a juice reward only at the end of each observation period (Figure 4). Once near-perfect performance was achieved, training was continued with stimuli that mimicked the stochastic changes of stimulus appearance during binocular rivalry. Simulating perception during rivalry with composed, mixed nonrivalrous stimuli allowed the monkey to grow accustomed to the "feel" of rivalry while still affording us full control over the animal's behavior, as incorrect responses would abort the observation period. Gradually, periods of real binocular rivalry were introduced into the emula-

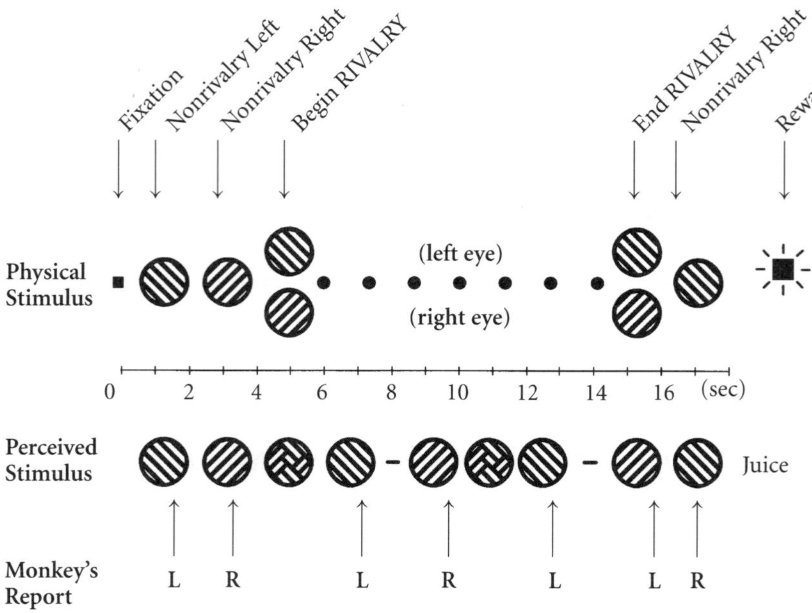

Figure 4. The monkey fixated a small spot and maintained fixation as a series of nonrivalrous (monocular or binocular) and rivalrous grating patterns were presented. The animal signaled perceptual transitions between the two orientations by pressing one of two levers (R or L) mounted on the primate chair. During the rivalry period the perceived transitions were not accompanied by changes in the physical stimulus. Incorrect responses for a nonrivalry trial, or failure to maintain accurate fixation (within a 0.8 degree window) resulted in abortion of the 15–25 second observation period, and the monkey would forfeit his juice reward.

tion, where the orientations in the right and left eyes were perpendicular. During these periods the monkeys continued to respond to orientation reversals, but now these changes were purely subjective as the physical stimuli remained unaltered, and hence feedback for inappropriate responses was impossible. The accuracy of response during these periods was, however, indirectly probed by introducing catch trials, in which the orientation of one of the gratings was smoothly replaced after a lever response to yield a coherent binocular stimulus. The monkey was expected to respond either immediately or not at all depending on whether the catch trial orientation was perpendicular to or the same as that indicated by the previous response, a test in which each monkey consistently performed above 95%.

In addition, two psychophysical controls were employed to show that the monkey was faithfully reporting his perception during rivalry. First, the distribution of dominance phase durations were compared to those obtained from humans under identical stimulus conditions. The normalized distribution of monkey's phases (Figure 3b) resembled that obtained from a human observer (Figure 3a), and both were representative of those described previously for both monkeys and humans (Myerson et al. 1981; Walker 1975). Even stronger evidence as to the reliability of the monkeys' reports came from the study of the effects of interocular contrast differences on the mean phase-duration. During rivalrous stimulation, increasing the stimulus strength in one eye increases the visibility of that stimulus, not by increasing its mean dominance phase, but by decreasing the mean period for which this stimulus remains suppressed (Fox & Rasche 1969; Levelt 1965). The data obtained from the monkey (Figure 3e) show the same relationship between stimulus strength and eye dominance as do the human data in the present (Figure 3d) and other studies. No random tapping of the levers could possibly yield this type of consistency, nor is it likely that animals or humans systematically adjust their behavior for different interocular contrasts.

During the behavioral-testing sessions single neurons were isolated in the central representations of the fourth visual area (V4) as well as the border of striate cortex and V2 (V1/V2). As the monkey fixated a small point, each individual cell was evaluated for its orientation selectivity and binocularity using a computer-automated procedure. Specialized glasses worn by the animal produced complete isolation of the right and left monocular images. Rivalry stimuli were constructed based on the preferred attributes of the specific cell, where the preferred orientation was placed in one eye and the orthogonal orientation in the other, whereas nonrivalry stimuli consisted of either the preferred or nonpreferred orientation presented either monocularly or binocularly. During testing, a typical observation period consisted of several nonrivalry periods, each lasting 1–5 seconds, and a single rivalry period, lasting up to 15 seconds (Figure 4). The monkey responded to orientation changes during the nonrivalry periods and to perceived orientation changes during rivalry, while individual neurons in his brain were monitored using standard extracellular recording techniques (see Leopold & Logothetis 1996). Of special interest were the rivalry periods, during which neurons exhibited a diversity of responses, and often modulated their activity according the monkey's perceived stimulus.

Figure 5 illustrates four representative examples of the cell types encountered. For each cell, the left and right plots represent the activity averaged over numerous perceptual transitions to the preferred and null orientations, respec-

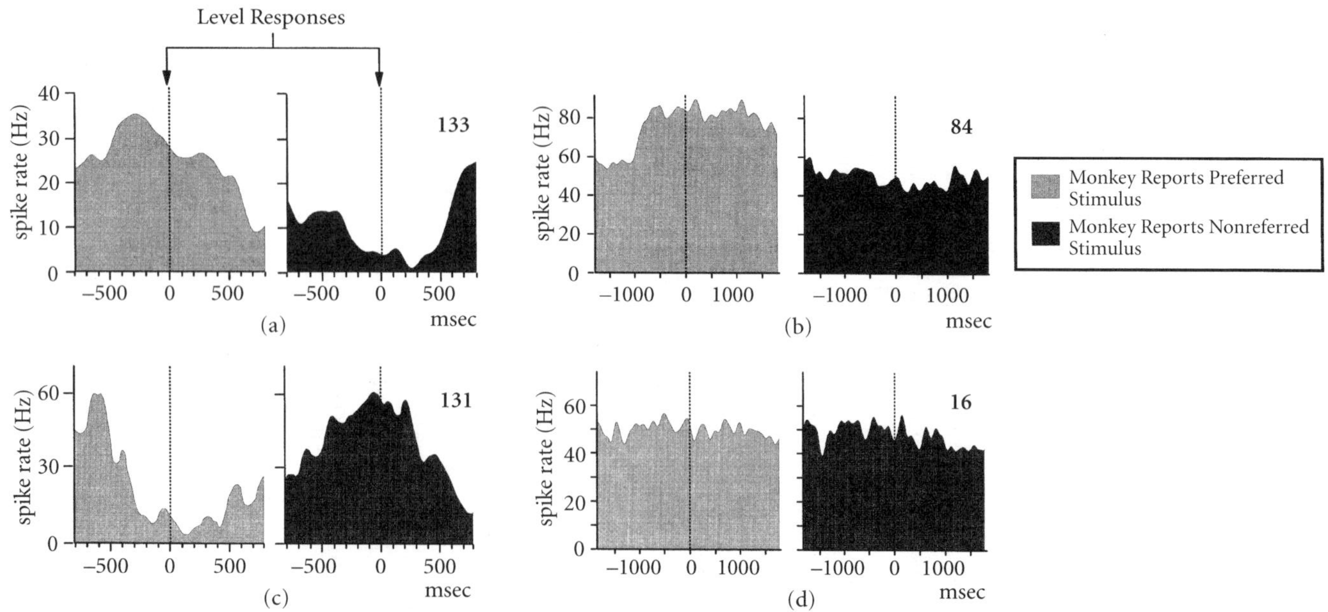

Figure 5. Four examples of the various cell types encountered during rivalry. For each pair of plots, the shaded regions correspond to the activity of the cell during rivalry around the time of a perceptual transition. The lighter shading represents trials in which the monkey reported a transition to the cell's preferred orientation, and the darker shading to the nonpreferred. The plots are centered around the animal's lever responses (vertical lines) and the activity is shown before and after the response. (a) The most common class of modulating cell, which increased its firing shortly before the monkey reported seeing a transition to the cell's preferred orientation and decreased before the nonpreferred orientation. (b) A cell whose elevation in activity was sustained as the monkey perceived the preferred orientation. (c) A cell which was representative of some cells in V4 that became more active when the monkey reported seeing the cell's nonpreferred orientation. (d) A non-modulating cell, whose activity was not influenced by the monkey's perceptual state.

tively. The activity is shown as a function of time, centered on the monkey's lever responses (vertical lines). Roughly one in three neurons tested modulated its activity in accordance with the perceptual changes. These modulating cells were almost exclusively binocular, and had a higher preponderance in area V4 (38%) than V1/V2 (18%). The majority of modulating cells exhibited an increase in their firing rate a fraction of a second before the monkey reported perceiving the cell's preferred orientation. Most often this elevation was temporary, declining again to the base rate after several hundred milliseconds (Figure 5a), although for some cells it was sustained significantly longer (Figure 5b). One class of neurons became most active when the nonpreferred stimulus was perceived, and the preferred stimulus was perceptually suppressed (Figure 5c). Finally, a large number of cells in all areas were generally uninfluenced by the animal's perceptual state (Figure 5d).

Cell responses in the inferior temporal cortex

Neurons in the visual areas of the anterior temporal lobe of monkeys exhibit pattern-selective responses that are modulated by visual attention and are affected by the stimulus in memory, suggesting that these areas play an important role in the perception of visual patterns and the recognition of objects (Desimone & Duncan 1995; Logothetis & Sheinberg 1996). We have therefore examined the activity of neurons in these areas during binocular rivalry (Sheinberg & Logothetis 1997). We have recorded from the inferior temporal cortex (IT) and from the lower bank of the superior temporal sulcus (STS). Because neurons in these areas respond better to complex patterns rather than oriented lines or gratings, the animal were taught to pull and hold the left lever whenever a sunburst-like pattern (left-object) was displayed, and to pull and hold the right lever upon presentation of other figures, including images of humans, monkeys, apes, wild animals, butterflies, reptiles, and various man-made objects (right-objects). In addition, the monkeys were trained not to respond or to release an already pulled lever upon presentation of a physical blend of different stimuli (mixed-objects). Examples of such objects can be seen in the insets of Figure 6.

Figure 6 shows two observation periods during this task, taken from two different monkeys. The rivalry stimuli (gray regions) were created by presenting the effective stimulus to one eye and the ineffective stimulus (the sunburst) to the other. Each plot illustrates the stimulus configuration, the neuron's activity, and the monkey's reported percept throughout the entire observation

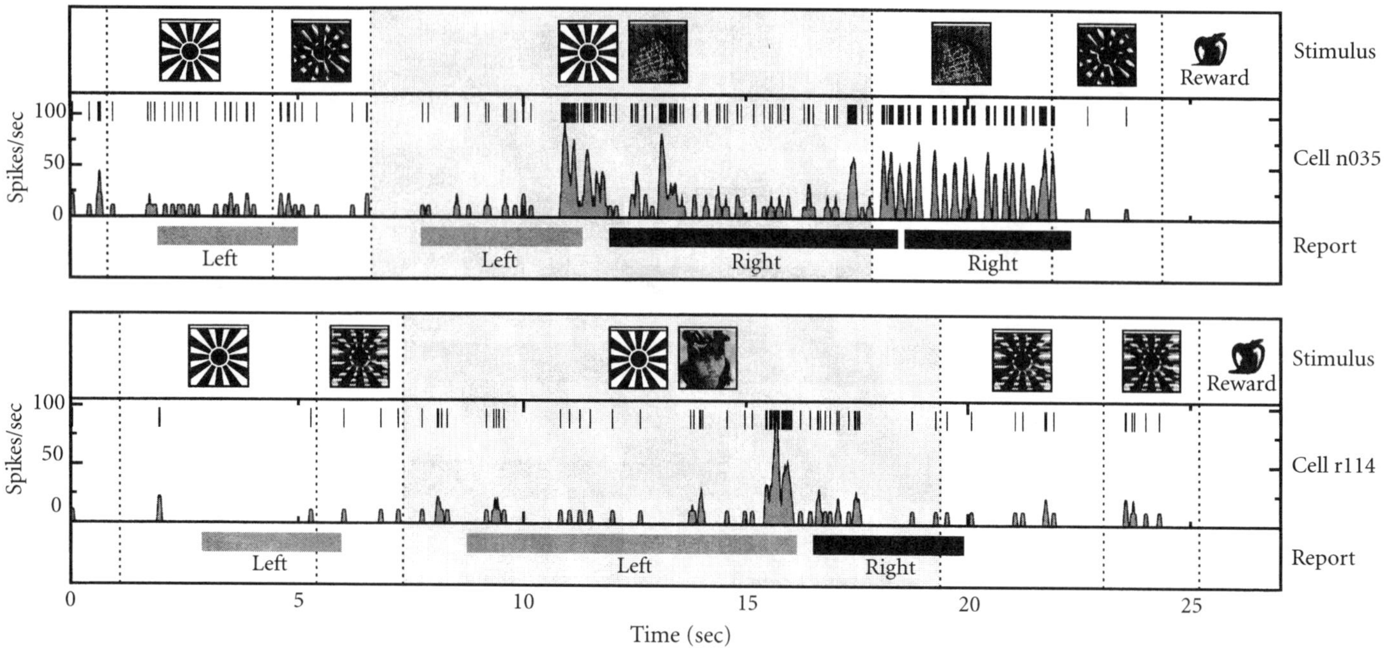

Figure 6. Behavioral and physiological response of IT/STS cells during rivaly observation periods. Panel depicts a single observation period, taken from different monkeys, during which non-rivalrous and rivalrous stimuli were present. Whenever the monkeys perceived seeing the sunburst pattern, they pulled the left lever, and when they perceived the complex object (roller coaster in (a) monkey face in (b)) they pulled the right lever. During rivalrous stimulation, both stimuli were presented, but the monkeys' percepts changed from seeing the sunburst pattern to seeing the effective stimulus. There is a clear correlation between the monkeys' reported change of percept and the cells' change in activity, even though the physical stimulus remain unchanged.

period. In each case, the neuron discharged only before and during the periods in which the monkey reported seeing the effective stimulus. During rivalrous stimulation, the stimulus configuration remained constant, but significant changes in cell activity were accompanied by subsequent changes in the monkeys' perceptual report. Of such selective neurons, 50 were tested during the object classification task under both non-rivalrous and rivalrous conditions.

To increase the instances of exclusive visibility of one stimulus, and to further ensure that the monkey's report accurately reflected which stimulus he perceived at any given time, we also tested the psychophysical performance of the monkeys and the neural responses of STS and IT cells using the flash suppression paradigm (Wolfe 1984). In this condition, one of the two stimuli used to instigate rivalry is first viewed monocularly for 1–2 seconds. Following the monocular preview, rivalry is induced by presenting the second image to the contralateral eye. Under these conditions, human subjects invariably perceive only the newly presented image and the previewed stimulus is rendered invisible. Previous studies have shown that the suppression of the previewed stimulus is not due to forward masking or light adaptation (Wolfe 1984) and that instead it shares much in common with the perceptual suppression experienced during binocular rivalry (Baldwin et al. 1996). In our experiments, the monkeys, just like the human subjects, consistently reported seeing the stimulus presented to the eye contralateral to the previewing eye during the flash suppression trials.

To confirm that the animals responded only when a flashed stimulus was exclusively dominant, catch trials were introduced in which mixed stimuli were flashed, after which the monkey was required to release both levers. Performance for both animals was consistently above 95% for this task. The upper plots of Figure 7 show the cell responses for monocular presentations, and the lower plots of the same figure show the neuron's activity at the end of the monocular preview (to the left of the dotted vertical line), and when perceptual dominance is exogenously reversed as the rival stimulus is presented to the other eye (to the right of dotted vertical line). The cell fires vigorously when the effective stimulus dominates perception and ceases firing entirely when the ineffective stimulus is made dominant. To better understand the differences between the temporal areas and the prestriate areas, recordings were also performed in area V4 using the flash suppression paradigm [Leopold & Logothetis, unpublished observations]. V4 neurons were significantly less affected by the perceptual changes during flash suppression. Presenting the ineffective stimulus after priming with the effective one caused little or no suppression of the discharge of any of the cells; presenting the effective stimulus after prim-

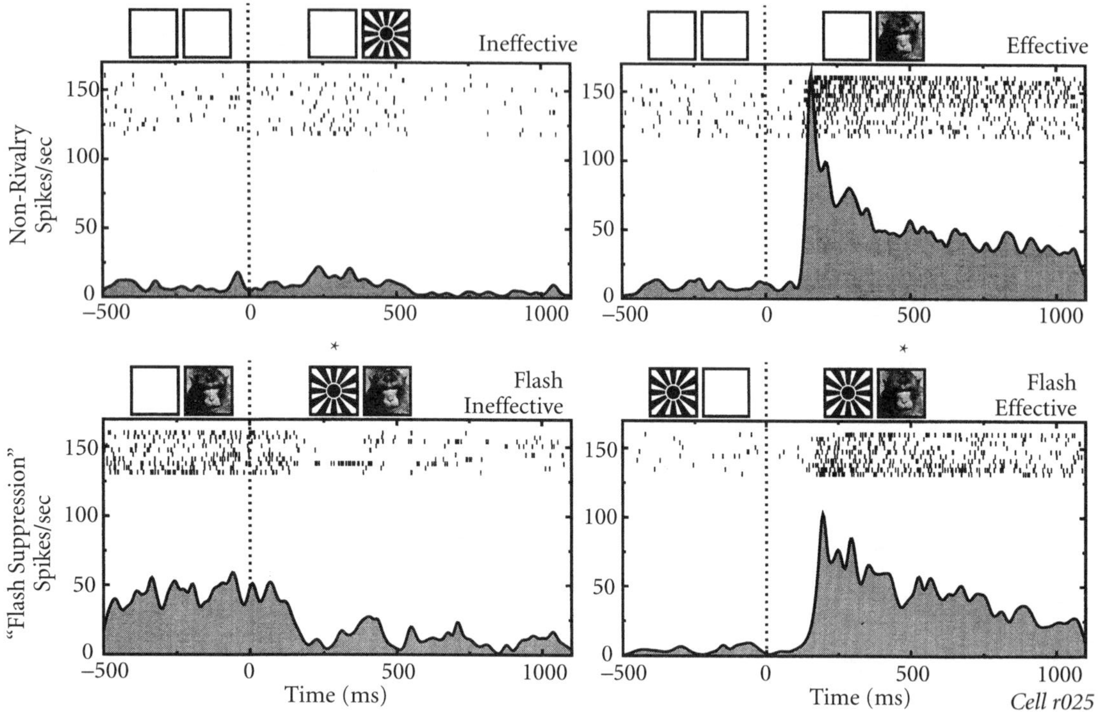

Figure 7. Example cell activity during presentations of the ineffective and effective stimuli during monocular presentations (a) and during the flash suppression paradigm (b), in which either the preferred stimulus or the non-preferred stimulus was first presented monocularly during a brief preview period, and then joined by the rival stimulus in the other eye. Just following the presentation of the new stimulus, the monkey reported seeing either the sunburst (bottom left) or the face (bottom right) exclusively. Likewise, the cell ceased firing when the sunburst was perceived but responded vigorously when the face was perceived, even though both stimuli were present in both conditions.

ing with the other had an weak effect on a small percentage of V4 neurons. The activity of the vast majority of studied temporal cortex neurons was found to be contingent upon the perceptual dominance of an effective visual stimulus. Neural representations in these cortical areas appear, therefore, to be very different from those in striate and early extrastriate cortex.

Neurons and perception

The studies described above make a few new points regarding both binocular rivalry and perception in general. First, the physiological results – just like the psychophysical results described earlier – are incompatible with the hypothesis that phenomenal suppression during binocular rivalry results from a blockade of information emanating from either eye. Eye-specific inhibition would be almost certainly reflected by decreased activity of monocular neurons (Blake 1989), yet most monocular cells remained entirely unaffected during rivalry suppression. Instead, the highest fraction of perception-related neurons were binocular and encountered in the extrastriate areas V4 (Leopold & Logothetis 1996) and MT (Logothetis & Schall 1989).

Second, it is interesting, though perhaps not surprising (given the responses obtained in the anesthetized preparation), that neural activity in visual cortex does not always predict awareness of a visual stimulus. While some neurons appear to modulate their activity in perfect correlation with the animal's changing perception, many others continue firing whether or not the stimulus is perceived. It will be of great interest to determine whether the different response properties of the response modulating neurons are in any way correlated with the anatomical location and connectivity patterns (Crick & Koch 1995).

Finally, the data presented here suggest (in agreement with a number of studies on the physiology of visual attention; for a review see Desimone & Duncan 1995) a prominent role of the extrastriate areas of the temporal stream in image segmentation and grouping. The early areas appear to have neurons that respond during perception but also during the suppression of the stimulus, and therefore appear to have the entire neural substrate of a system of reciprocal inhibition, explaining many of the characteristics of rivalry. They also have neurons, whose response depends on the task requirements, and whose activity is modulated by selective attention. Active inhibition, influences of attention, and selectivity of responses to complex two-dimensional patterns all strongly suggest an important role of early visual cortex in perceptual organization. In striking contrast with the early extrastriate areas, which show great variability in perception-related responses, in the inferior temporal cortex almost all neurons modulate their activity in synchrony with the monkey's perception when rivalry is initiated between faces or other complex images (Sheinberg & Logothetis 1997). The response variability in areas V2, V3, V4 and MT may be the result of the feedforward and feedback cortical activity that underlies the processes of grouping and segmentation – processes that are probably perturbed when viewing ambiguous figures. The consistent correlation of cell activity with perceptual report in the areas of the temporal lobe may, on the other hand, represent a stage of processing beyond the resolution of perceptual ambiguities, where, neural activity reflects the integration of constructed visual percepts into those subsystems responsible for object recognition and visually guided action.

In the introductory section we asked: Which cells code for the "hodgepodge" of visual elements and which related directly to our knowledge of familiar objects? Obviously our experiments do not purport to answer this question, nor did we expect to understand perceptual organization in a few experiments or by only studying single cells and examining their average rate of firing. The study of dynamic interactions among neurons within and between areas (for review see Singer & Gray 1995) will be of great importance for understanding image segmentation, as will also be the identification of different types of modulating neurons and their connectivity. Combination of such techniques in experiments with alert, trained animals using stimuli that instigate perceptual multistability may help us gain important insights into the neural processes that underlie the conscious perception of a visual stimulus.

References

Baldwin, J. B., Loop, M. S., Edwards, D. J. (1996). Magnitude and time course of interocular suppression is stimulus selective. *Investigative Ophthalmology and Visual Science Supplement, 37* (3), 3016.

Blake, R. R. (1989). A Neural Theory of Binocular Rivalry. *Psychological Review, 96*, 145–167.

Borsellino, A., De Marco, A., Allazetta, A., Rinesi, S., & Bartolini, B. (1972). Reversal time distribution in the perception of visual ambiguous stimuli. *Kybernetik, 10*, 139–144.

Campbell, F. W., Howell, E. R. (1972). Monocular alternation: A method for the investigation of pattern vision. *J. Physiol. (Lond.), 225*, 19P–21P.

Crick, F., & Koch, C. (1995). Are we aware of neural activity in primary visual cortex. *Nature, 375*, 121–123.

Desimone, R., & Duncan, J. (1995). Neural mechanisms of selective visual attention. *Annual Review of Neuroscience, 18*, 193–222.

DuTour, M. (1760). Discussion d'une question d'optique (discussion on a question of optics) Academie des Sciences. Memoires de Mathematique et de Physique Presentes par Divers Savants.

Fox, R., & Herrmann, J. (1967). Stochastic properties of binocular rivalry alternations. *Perception and Psychophysics, 2*, 432–436.

Fox, R., & Rasche, F. (1969). Binocular rivalry and reciprocal inhibition. *Perception and Psychophysics, 5*, 215–217.

Lathrop, R. G. (1966). First-order Response Dependencies at a Differential Brightness Threshold. *J. Exp. Psychol., 72*, 120–124.

Leopold, D. A., Fitzgibbons, J. C., & Logothetis, N. K. (1995). The role of attention in binocular rivalry as revealed through optokinetic nystagmus. *Artificial Intelligence Memo, 1554, C.B.C.L #126*, l–17.

Leopold, D. A., & Logothetis, N. K. (1995). Cell Activity Reflects Monkeys' Perception During Binocular Rivalry. *Invest. Ophthalmol. Vis. Sci. (Suppl.), 36*, S813.

Leopold, D. A., & Logothetis, N. K. (1996). Activity changes in early visual cortex reflect monkeys' percepts during binocular rivalry. *Nature, 379*, 549–553.

Levelt, W. J. M. (1965). On Binocular Rivalry. In *Vision: Binocularity and Binocular Depth* (pp. 1–110). Assen: Royal VanGorcum Ltd.

Logothetis, N. K., & Leopold, D. A. (1995). On the Physiology of Bistable Percepts. *Artificial Intelligence Memo No.*, 1–20.

Logothetis, N. K., Leopold, D. A., & Sheinberg, D. L. (1996). What is rivalling during binocular rivalry? *Nature, 380*, 621–624.

Logothetis, N. K., & Schall, J. D. (1989). Neuronal correlates of subjective visual perception. *Science, 245*, 761–763.

Logothetis, N. K., & Sheinberg, D. L. (1996). Visual object recognition. *Annual Review of Neuroscience, 19*, 577–621.

Myerson, J., Miezen, F., & Allman, J. (1981). Binocular Rivalry in Macaque Monkeys and Humans: A Comparative Study in Perception. *Behaviour Analysis Letters, 1*, 149–159.

Sheinberg, D. L., & Logothetis, N. K. (1997). The Role of Temporal Cortical Areas in Perceptual Organization. *Proc. Natl. Acad. Sci.*

Walker, P. (1975). Stochastic properties of binocular rivalry alternations. *Perception and Psychophysics, 18*, 467–473.

Wolfe, J. (1984). Reversing Ocular Dominance and Suppression in a Single Flash. *Vision Research, 24*, 471–478.

CHAPTER 7

Attention versus consciousness

A distinction with a difference

Valerie Gray Hardcastle
Virginia Tech, Blacksburg

What must be admitted is that the definite images of traditional psychology form but the very smallest part of our minds as they actually live. The traditional psychology talks like one who should say a river consists of nothing but pailsful, spoonsful, quartpotsful, barrelsful, and other moulded forms of water. Even were the pails and the pots all actually standing in the stream, still between them the free water would continue to flow. It is just this free water of consciousness that psychologists resolutely overlook. Every definite image in the mind is steeped and dyed in the free water that flows round it. With it goes the sense of its relations, near and remote, the dying echo of whence it came to us, the dawning sense of whither it is to lead. The significance, the value, of the image is all in this halo or penumbra that surrounds and escorts it, – or rather that is fused into one with it and has become bone of its bone and flesh of its flesh; leaving it, it is true, an image of the same thing it was before, but making it an image of that thing newly taken and freshly understood (James 1892: 165–166).

1. Introduction: James's legacy

Psychologists and neuroscientists alike often mention William James with approval as one who understood early on the fundamental connection between attention and consciousness. They point out that James was the first to see very clearly that we are only conscious of a tiny subset of the total of our "impressions from our whole sensory surface" (James, p. 217), and they go on to assume both that attention and consciousness amount to the same thing and that that is what James believed (see, e.g., Crick 1994; LaBerge 1995; Newman 1997a). Both assumptions are in fact mistaken. Attention is importantly different from consciousness and learning about the former should not be confused with learning about the latter. And though James did appreciate the importance of attentional processing in cognition, he did not hold that what we pay attention to exhausts conscious experience. Indeed, as the quotation above illustrates, he believed that defining consciousness in terms of some other process is much too narrow and is a mistake on psychology's part.

It is, unfortunately, a mistake we are still making (see Baars 1988; Crick & Koch 1990; He, Cavanaugh, & Intriligator 1996; Newman 1995, 1997b; Posner 1994; Strehler 1991) and I shall advocate in this paper differentiating attention from consciousness. In particular, I believe that James's view is correct: our experience contains both items we pay attention to and things in the background which connect, fill out, and enhance what lies in the focus of our attention. Before I launch into my arguments, however, I shall outline some reasons why psychologists and neuroscientists might conclude that attention and consciousness are the same thing. I do not believe that such a conclusion is unreasonable; only that it is false.

I should also note at the outset that I shall be taking reportability (broadly construed) as a useful criterion in determining whether someone is having a conscious experience. This move should not be read as assuming that reportability is necessary or sufficient for consciousness – it isn't – but it is one good way to gather data to hypothesize about the internal states of others. I am not going to defend or discuss my methodology here; it has been discussed at length elsewhere (cf., Dennett 1991; Flanagan 1992).[1]

2. The link between attention and consciousness

We can find considerable telling evidence suggesting that we are only conscious of items to which we are attending. Here I rehearse three significant results from psychology and clinical neurology that supports the general position that attention is fundamentally tied to, if not identical with, consciousness.

2.1 Confabulation across saccades

Our impression in looking at some complex scene is of a smooth panorama. We are generally unaware of our eyes darting back and forth across our visual world. Nevertheless, we look by fixating on some object or feature before us for about 1/4 of a second, then quickly moving the center of our gaze elsewhere. These saccades are quite rapid, usually lasting about 20 msec or so, before the next pause. The reason we have difficulty sensing our rapid eye movements is that we are effectively blind during them. Our visual world is actually a gappy one of brief and disjointed "snapshots" of different parts of our environment interspersed with brief interludes of no visual inputs at all. Obviously our brains fill in what we do not see to give us the experience of a continuous and stable scan.

An interesting consequence of our brain's drive to hide our visual lapses is an inability to remember from one fixation to the next. For example, when subjects read sentences written in alternating case that reverses every saccade they are apparently unaware of the changes (McConkie & Zola 1979). Subjects would have sentence (1) below before them on a viewing screen. As they began an eye movement to another location on the screen, sentence (1) would be exchanged for sentence (2). However, they would report nothing unusual happening and the change would not interfere with their reading patterns.

(1) YoU cAn FoOl SoMe Of ThE pEOPlE sOmE oF tHe TiMe BuT yOu CaN't FoOl AlL tHe PeOpLe AlL tHe TiMe.

(2) yOu CaN fOOL sOmE oF tHe PeOpLe SoMe Of ThE tImE bUt YoU cAn'T fOoL aLl ThE pEOPlE aLl ThE tImE.

Indeed, when David Zola first tested the equipment for the experiment, he reported that the eye tracking apparatus must be broken since nothing happened with the sentences, even though the changes were perfectly obvious to the bystanders in the room who were not having the alternations keyed to their saccades (as reported in Grimes 1996). Similar experiments come to the same conclusions. When text is shifted to the left or right during a saccade so that the eye would focus on a different letter than the one originally intended, subjects would make a small saccade to compensate for the moved words. However, they would rarely report any awareness of the change (Grimes 1996). When text outside the focus of attention is replaced with nonsense shapes, subjects are unaware that they are not viewing an entire sentence. Our "window" of veridical conscious perception extends only about fifteen characters to the right and four characters to the left of the point of fixation when reading right to left (McConkie & Rayner 1975, 1976; Rayner & Pollatsek 1987; Rayner, Well, & Pollatsek 1980; Underwood & McConkie 1985). (The effect follows the direction of reading (Osaka & Oda 1991; Pollatsek, Bolozky, Well, & Rayner 1981).)

The lack of conscious access to peripheral information across saccades extends beyond linguistic stimuli. When subjects scan a picture, they are quite often unaware of changes made in the objects in the scene if the changes are made during eye movements. Subjects can be told that some objects might shift while they are looking at the pictures and to indicate immediately when this happens. The changes can be quite dramatic and can even occur with objects that a subject has just fixated upon. Still they remain unnoticed. For example, a prominent building in a city skyline became 25% larger and 100% of the subjects failed to detect any change. One hundred percent also failed to notice that two men exchanged hats of different colors and styles. Ninety-two percent did not detect that a third of a crowd of 30 puffins disappeared. Eighty-three percent did not see that in a playground scene, a child is moved forward about five meters and enlarged by 30%. Most strikingly, 58% of subjects did not notice that the swimsuit of one of four people in an advertisement for a swimsuit changed from bright pink to bright green, even though they began viewing the picture fixated on the green suit (all these results are reported in Grimes 1996).

These subjects ignored or were insensitive to new visual information right under their noses, as it were. How could this be happening? Clearly our visual experiences are not based upon the details of the visual scene. Instead, we construct what we assume must be out there based on our rough and ready impressions derived from information we get when we pay attention to portions of the world before us. Apparently people have very little conscious access to peripheral information received across saccades. What occurs outside the focus of attention is either lost or is never really processed in the first place. All people can report accurately is what currently lies in the focus of attention; it seems that everything else is an educated but confabulated guess about what was probably there outside of awareness.

2.2 Metacontrast masking

The phenomenon of backward masking or metacontrast has been known and studied for quite some time. If a solid figure or word is briefly shown on a screen, followed by a different figure, subjects report only the second object. The information from the first presentation is simply not seen; it is masked by what follows. Psychological experiments using metacontrast tell us much about the time course and steps of visual processing.

However, it is now becoming clear that we can manipulate the effects of backward masking using attention. We lose the masking effect of the metacon-

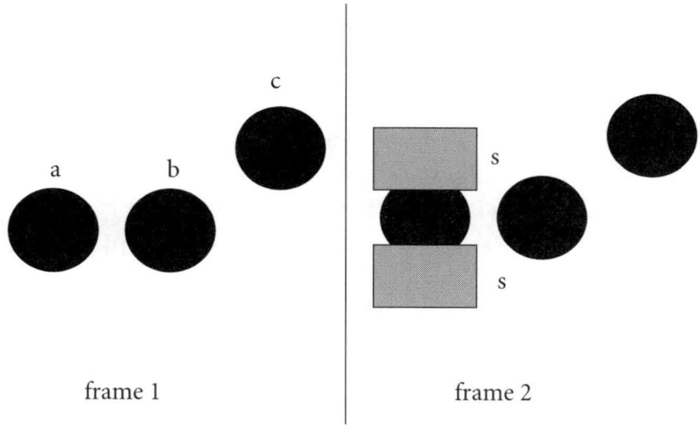

Figure 1. Metacontrast display demonstrating attentional effects. The target is disk a. Disks b and c are shown simultaneously with a. Frame 1 is replaced by frame 2, which includes the two rectangles marked s. When subjects paid attention to a and b as a unit, a was not masked by s. When subjects grouped b and c, a was not reported [after Ramachandran & Cobb 1995].

trast stimuli if we do not pay attention to the target (Raymond, Shapiro, & Arnell 1992). Moreover, when a target circle is perceptually grouped with another circle next to it, we lose the masking effects of two flanking rectangles. But if the target circle is not grouped with the second circle – even though the second circle still appears in the same place on the screen, but is now grouped with a third circle – the masking effect returns (Ramachandran & Cobb 1995). (See Figure 1.) This effect can be repeated using Gestalt groupings as well. If we see the target as part of a row or column of flanking circles, then it isn't masked, but if the target falls outside the Gestalt grouping, it is masked.

Here as before, we find a fairly direct connection between attention and our conscious experiences. By focusing our attention differently, we are able to see different stimuli configurations among the same set of inputs. And in doing so, we can make the target circle seem to disappear and reappear, almost at will. How we pay attention to the world seems to be tightly correlated with how and whether the world appears to us.

2.3 Disorders of attention

Without going into the details of the neurophysiology of visual attention (though see Desimone & Duncan 1995; LaBerge 1995, for excellent reviews), I would like to mention two attention-based deficits in cognition relevant to the discussion here: right parietal extinction and Balint's syndrome. If the right parietal lobe is lesioned, patients often exhibit some form of neglect of the contralateral visual field. In particular, they ignore stimuli occurring on their left sides, and when describing visual images, they omit the details of the left sides of what they visualize.

In addition, patients with right parietal lesions exhibit a form of visual extinction. When shown two objects, one contralateral and one ipsilateral to the lesioned hemisphere, subjects will report seeing only the one in the ipsilateral field. In some cases, if patients are shown more than one object to their intact hemifield, they will fail to see the leftmost one (Rapcsak et al. 1995; Di Pellegrino & De Renzi 1993). Somehow, processing the object on the right extinguishes awareness of the one on the left.

This sort of neglect is thought to be attentional and not visual because patients can still see objects in front of their damaged hemisphere if there is not anything being shown to the undamaged side that could capture visual attention. Hence, the visual paths themselves are at least only minimally damaged. Moreover, patients can see what is located contralaterally if their attention is inadvertently directed into that visual space or if they are told to ignore the stimuli in the ipsilateral field (Di Pellegrino & De Renzi 1993; Kamath 1988; Pizzamiglio, Guariglia, Incoccia, & Antonucci 1990; Rapcsak, Watson, & Heilman 1987; Robertson 1989; Rubens 1985; Walker, Findlay, Young, & Welch 1991). This tells us that right parietal patients lose perception in their left side when objects there must compete with those on the right for visual resources. The lesions seem to prevent patients from moving their focus of attention to their left if it is already captured on the right and, as a result, they cannot see or report what occurs there. Once again, we find that a lack of attention means a lack of consciousness.

If both parietal lobes are damaged instead of just one, then we get simultagnosia or Balint's syndrome (Balint 1909). In this fairly rare disorder, patients report being able to see only one object in visual space. They cannot integrate information across their visual fields. Instead, it seems that they can only pay attention to one small area at a time, see what is there, and then move to the next area. Patients lose all object or semantic information outside a tiny sweep of attention; the rest of the visual space is extinguished or suppressed. Given the phenomenon of right parietal extinction, simultagnosia should not be a surprise, as it is a more extreme version of extinction. But for a tiny window of object awareness, all information in the visual field is ignored or omitted. All patients can report seeing successfully is a small, restricted area; the rest is an undifferentiated mess.

This brief discussion indicates that we do not perceive much information outside our focus of attention, that what we do see is quickly lost, and that voluntarily directing attention controls the appearance and disappearance of perceptual objects. It seems clear that attentional processing plays a decisive role in our conscious perceptions, so it is unsurprising that Michael Posner concludes in his review of attention and consciousness that "an understanding of consciousness must rest on an appreciation of the brain networks that subserve attention, in much the same way as a scientific analysis of life without consideration of the structure of DNA would seem vacuous" (1994:7398).

3. The distinction between attention and consciousness

I admit the data from the above section present a compelling and simple story. It would be so nice if it were true, too, for identifying attention and consciousness would give us not only a well-developed research program for studying awareness but also lots of information about consciousness itself gleaned from previous studies of attention. However, the relation between attentional processing and phenomenal experience is more complicated and subtle. Each of the analyses presented above admits of alternative interpretations. In this section I shall discuss different ways of approaching attention and consciousness as well as present additional evidence in support of maintaining the distinction between them.

3.1 Saccadic blindness as failure to remember

I mentioned in the introduction that I would be taking verbal reports of some phenomenal experiences as strong evidence for consciousness, for this is the best criterion we currently have. However, this method is not foolproof; moreover, our best method is neither perfectly straightforward nor wildly successful. In particular, it is most important not to confound reports of conscious states with mnemonic reports. Our normal confabulations to explain what we miss across saccades and what we thought we saw in previous fixations could reflect merely an inability to report what we were just aware of. It could be that we see items at the periphery consciously, but that information does not make it into short term store or some other memory system so we are prevented from recalling our earlier experiences a short while later.

We have known for almost forty years, since Sperling's seminal work on partial reports, that subjects perceive more than they can report (Averbach

& Sperling 1968; Sperling 1960). Items not explicitly reported are available in "iconic" storage for a few hundred msecs. Sperling believed is that all incoming visual information first passes through a pre-categorical memory. To be reported as a visual experience, the information must be converted into a different form. Attention was thought to do the converting; subjects shifted attention from one perceptual item to the next, reading them off as semantic objects until the information could no longer be maintained in iconic storage.

Most researchers assume that iconic storage is largely unconscious, but that belies subjects' reports. Sperling himself embarked on his investigation because subjects so often complained that they could see more than they could report. How could people be able to note this fact unless they experienced a substantial portion of the visual array, albeit in perhaps a vague and inchoate way? I submit that they could not, and that more visual information is consciously perceived than can be explicitly listed or noted. (Iwasaki 1993, presents roughly the same argument.)

Sperling's original hunch that we can consciously perceive items that lie outside our focal attention dovetails with recent results from psychological experiments which explore exactly what we can see in the periphery. It is clear that we can detect some stimuli in our periphery when we are focused on foveal targets, especially if the peripheral information has a similar spatial frequency and orientation to the central objects of perception (Rossi & Paradiso 1995). We can see textual discontinuities without attention, though we seem to need attentional processes to perceive Gestalt groupings (Ben-Av, Sagi, & Braun 1992). We can perceive shape-from-shading pop-outs independently of attention (Braun 1993). Finally, peripheral objects can be seen and reported when they are embedded in textual noise (Braun & Sagi 1990), and we are better at reporting low spatial frequency information if it is outside our focus of attention (Shulman & Wilson 1987; Wong & Weisstein 1983).[2]

It is also clear that what we perceive in the periphery is far less than what we see using focal attention. But, the reason why our perception outside of attention is relatively poor remains as of yet undetermined. The inhomogeneity of the retino-striate system explains a large portion of the inferiority of perception outside the fovea (Saarinen 1993). Since attention is generally directed toward foveal information, it is hard to sort out whether superior processing of focal information is due primarily to attention itself or is to fewer brain resources being devoted to extrafoveal processing in general.

That subjects cannot accurately report items appearing at a previous point of fixation does not tell against unattended information being conscious. Since we can consciously see more than we can report and at least some peripheral stimuli, at best it suggests that many conscious experiences are not preserved for very long.

3.2 Metacontrast as motion capture

A different explanation for metacontrast phenomena – and one endorsed by Ramachandran and Cobb (1995) – is that it stems from the same mechanisms that produce apparent motion. Since the target and the mask occur one after the other, the brain tries to read that information as the target moving to the masking stimuli's location(s). Consequently, we are no longer able to see the target at its original location since objects are not supposed to be in two places at the same time. In support of this hypothesis, Ramachandran and Cobb note that subjects in their backward masking experiments sometimes report the target circle seeming to split into the masking rectangles. In addition, when subjects viewed a display with two rectangles followed by the two masking rectangles next to them, they vividly saw the shapes move horizontally, and did not see the target split into being rectangles, nor was the target masked. (See Figure 2.) The apparent motion of the masking rectangle trumped their masking effects. If Ramachandran and Cobb's hypothesis is correct, then the reason we do not perceive the target under backward masking conditions is that we have confused our motion detectors. Paying attention to the target or surrounding stimuli per se might have little directly to do with our awareness of backwardly masked figures.[3]

Figure 2. Preferential perception of apparent motion. When the two rectangles appear to move hoizontally from frame 1 to frame 2, the target circle is not masked [after Ramachandran & Cobb 1995].

Nevertheless, attention does have a great deal to do with exactly how stimuli appear to us. Attentional processes can alter the content or flavor of our conscious perceptions. They affect the perceived brightness and subjective length of stimuli, the organization of ambiguous figures, as well as other similar factors (Tsal 1994; Tsal, Machshon, & Lamy 1995). For example, directing attention to a line gives us the impression that it is being drawn across our visual field, even though it is presented all at once (Hikosaka, Miyauchi, & Shimojo 1993).

In addition, some hold that we cannot process the meaning of text without attention, though we can process its form (Driver & Baylis 1991; Inhoff & Briihl 1991; Mulligan & Hartman 1996; Posner, Sandon, Dhawan, & Shulman 1989; though see Berti, Frassinetti, & Umilta 1994). However, I believe that the relation between the meaningfulness of stimuli and attention follows from attention's close tie to explicit memory. We now have substantial psychological and neurophysiological evidence that attention and explicit memory are strongly correlated (Craik, Govoni, Naveh-Benjamin, & Anderson 1996; Fuster 1990; Grimes 1990; Kellog, Newcombe, Kammer, & Schmitt 1996, Richardson-Klavehn, Gardiner, & Java 1994; Szymanski, & MacLeod 1996). That is, paying attention during learning is important (though not necessary) for later remembering events explicitly and consciously, but it is not required for remembering something implicitly or without awareness. I have argued at length elsewhere that explicit memory is what allows our interpretations of incoming stimuli to appear richly meaningful and gives them the particular qualitative feel that they have (Hardcastle 1995). If I am correct, and I shall not rehearse my reasons for my view here, then, again, attention per se does not allow for our conscious perceptions. Instead, attention's connection to other processes gives the appearance of attention causing awareness.

On the other hand, if I am wrong and attention is in fact directly responsible for stimuli appearing meaningful to us, then this fact would still not show that attention is co-extensive with consciousness. Indeed, some process affecting the shape of a qualitative experience indicates that the process is not identical to consciousness. Attention may allow us to process stimuli more deeply or more fully. It may even allow us to process information differently. But none of this shows that by doing so we *ipso facto* get consciousness.

3.3 Assuming what we want to prove

Assessing what lesion patients' qualitative experiences are like is a difficult proposition, especially if we agree that subjects can often see more than they can report. Direct questioning does not always produce the full phenomenal story, and parietal patients are no exception. When explicitly asked, "Do you know what this is?" they are unable to identify neglected objects or words. However, in a forced-choice matching task, they can retrospectively confirm the accuracy of their response. "Are you sure?" is met with vigorous assent (Feinberg, Dyckes-Berke, Miner, & Roane 1995).

Is it the case that parietal patients see nothing in their neglected areas or is it that they have trouble integrating what they see or assigning it semantic content or verbalizing their elusive experiences? It is hard to know for sure. Certainly, patients with Balint's syndrome are aware of more things in their visual field than the single object they can identify. They see an undifferentiated mess, to be sure, but they are seeing nonetheless.

How should we interpret the extinction phenomena? Under certain conditions, it is possible for patients with hemineglect to see and report objects in the contralesional field even though items are also present in the ipsilesional field. For example, if nonsense lines are presented to the undamaged hemisphere, but a contentful object is shown to the lesioned side, patients report seeing what is contained in their neglected space (Ward & Goodrich 1996). In addition, when the objects across the visual fields create a subjective surface or complete an occluded object, neglect patients accurately report both the surface and its components or the partially obscured object (Mattingley, Davis, & Driver 1997). (See Figure 3.) Attention is an object-oriented, surface-based function, while pre-attentive processes create the surfaces for attention to manipulate. Objects – illusory or otherwise – on the neglected side can capture attention if they have few rivals. But these facts tell us little about the actual experiences of parietal patients. It is difficult to report unattended stimuli, even without brain damage, but without already assuming that attention and consciousness are intimately connected, we cannot use the lesion data to conclude that an inability to note stimuli explicitly means a nonqualitative unawareness of the input. These studies tell us no more than that extinction is a problem with directing attention and that attention operates on objects not features. With parietal patients able to re-identify with confidence shapes they were shown and knowing that we can see at least low level features outside of attention under normal conditions, we should conclude that parietal patients most likely can see the peripheral stimuli in some fashion as well. I suspect they have difficulty retaining the inputs in memory and consequently attaching meaning to them, but that is a different story.

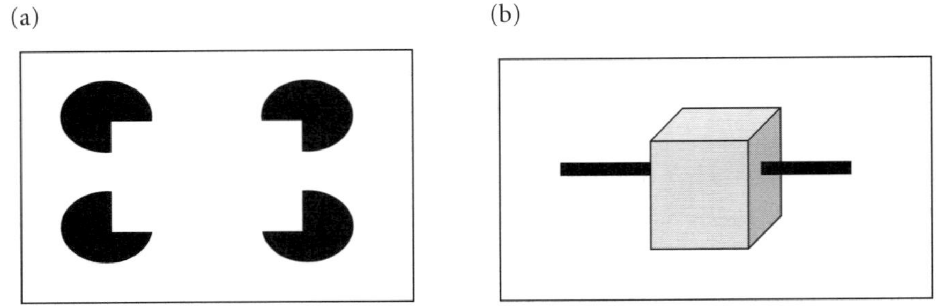

Figure 3. Examples of displays that extinction patients do not neglect. In both cases, the information contained on the contralesional visual field forms part of a subjective whole with the stimuli presented on the ipsilesional side. In (a), the Pac-men create the illusion of a square. In (b), the bar appears occluded by the cube [derived from Mattingley et al. 1997].

4. The neural correlates of consciousness: A non-answer

Thus far, I have argued for a purely negative thesis: we should not pretend that attention and consciousness amount to the same thing. This tells us little about consciousness, its neural correlates, or its function. Unfortunately, I have little more to add to this anti-position.

We know approximately the areas of sensory cortex that become active when we fill in surfaces, derive shape from shading, or otherwise analyze visual features. These are also the same areas that are active when we process information automatically (Posner 1994). What happens differently there with conscious awareness? I have argued that it is not the addition of attention.[4] It might still be activation of some form of explicit memory (cf., Hardcastle 1995; Moscovitch 1995), but what that means in this context is quite murky. James is probably right that "the cortex is the sole organ of consciousness in man (1890: 66; see also Bottini, Paulesu, Warburton, Wise, Vallar, Frackowiak, & Frith 1995), but it is also the organ of much unconscious processing, too. How to distinguish James's riverbed for consciousness from the bank on either side remains a deep mystery .. and an exciting but still unanswered intellectual challenge.[5]

Notes

1. This essay will also assume some understanding of what is meant by attention and consciousness, at least metaphorically. This is a highly dubious assumption, but space limitations prevent of fuller discussion.

2. Ned Block (personal communication) suggested that we might be able to spread our attention across the entire visual field so that peripheral experience would not tell against identifying consciousness with attention. However, given what we know about cellular effects of attention in inferotemporal and parietal cortex (cf., Desimone & Duncan 1995), this view is untenable. Neurons there respond differentially to a chosen subset of visual stimuli, processing these data at the expense of others. Our best evidence suggests that attention cannot expand to contain our entire visual world.

3. This is not Ramachandran and Cobb's interpretation of their own hypothesis. I do not mean to imply in my discussion that they would support my analysis of the relation between attention and metacontrast. Bruce Bridgman (personal conversation) believes that the motion capture view of metacontrast does not cover all cases of backward masking, though neither of us could come up with definitive examples one way or the other. He also reminded me that Daniel Kahneman first put forth the motion detection hypothesis.

4. Braun and Sagi (1990), Crick and Koch (1990), and Iwasaki (1993) argue that there are two forms of consciousness: attentional and peripheral or iconic. But without any serious evidence to support different types of consciousness as opposed to attentional versus non-attentional processing, parsimony demands that we assume one form of consciousness and that attention changes the content of our conscious states.

5. An earlier version of this paper was presented at the First Conference for the Society for the Scientific Study of Consciousness. My thanks to the participants for their lively and enthusiastic conversations. In particular, I owe thanks to Ned Block, Bruce Bridgeman, Cristof Koch, Bruce Mangan, and Iyal Reingold. This article appeared in *Cognitive Studies: Bulletin of the Japanese Cognitive Science Society* (1997), *4*, 56–66.

References

Averbach, E., & Sperling, G. (1968). Short term storage of information in vision. In R. H. Haber (Ed.), *Contemporary Theory and Research in Visual Perception* (pp. 202–214). New York: Holt, Rinehart, and Winston.

Baars, B. J. (1988). *A Cognitive Theory of Consciousness*. Cambridge: Cambridge University Press.

Bálint, R. (1909). Seelenlämung des "Schauens", optische Ataxie, räumliche Störung der Aufmerksamkeit. *Monatschrift für Psychiatrie und Neurologie*, *25*, 51–81.

Ben-Av, M. B., Sagi, D., & Braun, J. (1992). Visual attention and perceptual grouping. *Perception and Psychophysics*, *52*, 277–294.

Berti, A., Frassinetti, F., & Umilta, C. (1994). Nonconscious reading? Evidence from neglect dyslexia. *Cortex*, *30*, 181–197.

Bottini, G., Paulesu, E., Sterzi, R., Warburton, E., Wise, R. J., Vallar, G., Frackowiak, R. S., & Frith, C. D. (1995). Modulation of conscious experience by peripheral sensory stimuli. *Nature, 376*, 778–781.

Braun, J. (1993). Shape-from-shading is independent of visual attention and may be a "texton". *Spatial Vision, 7*, 311–322.

Braun, J., & Sagi, D. (1990). Vision outside the focus of attention. *Perception and Psychophysics, 48*, 45–58.

Craik, F. I., Govoi, R., Naveh-Benjamin, M., & Anderson, N. D. (1996). The effects of divided attention on encoding and retrieval. *Journal of Experimental Psychology: General, 125*, 159–180.

Crick, F. (1994). *The Astonishing Hypothesis*. New York: Macmillian.

Crick, F., & Koch, C. (1990). Towards a neurobiological theory of consciousness. *Seminars in Neuroscience, 2*, 263–275.

Dennett, D. C. (1991). *Consciousness Explained*. Cambridge, MA: The MIT Press.

Desimone, R., & Duncan, J. (1995). Neural mechanisms of selective visual attention. *Annual Review of Neuroscience, 18*, 193–222.

Di Pellegrino, G., & De Renzi, E. (1993). *Neuropsychologia, 33*, 153.

Driver, J., & Baylis, G. C. (1991). Target-distracter separation and feature integration in visual attention to letters. *Acta Psychologica, 76*, 101–119.

Feinberg, T. E., Dyckes-Berke, D., Miner, C. R., & Roane, D. M. (1995). Knowledge, implicit knowledge and metaknowledge in visual agnosia and pure alexia. *Brain, 118*, 789–800.

Flanagan, O. (1992). *Consciousness Reconsidered*. Cambridge, MA: The MIT Press.

Fuster, J. M. (1990). Inferotemporal units in selective visual attention and short-term memory. *Journal of Neurophysiology, 64*, 681–697.

Grimes, J. (1996). On the failure to detect changes in scenes across saccades. In K. Akins (Ed.), *Perception* (pp. 89–110). New York: Oxford University Press.

Grimes, T. (1990). Audio-video correspondence and its role in attention and memory. *Educational Technology Research and Development, 38*, 15–25.

Hardcastle, V. G. (1995). *Locating Consciousness*. Amsterdam: John Benjamins Press.

He, S., Cavanaugh, P., & Intriligator, J. (1996). Attentional resolution and the locus of visual awareness. *Nature, 383*, 334–337.

Hikosaka, O., Miyauchi, S., & Shimojo, S. (1993). Focal visual attention produces illusory temporal order and motion sensation. *Vision Research, 33*, 1219–1240.

Inhoff, A. W., & Briihl, D. (1991). Semantic processing of unattended text during selective reading: How the eyes see it. *Perception and Psychophysics, 49*, 289–294.

Iwasaki, S. (1993). Spatial attention and two modes of visual consciousness. *Cognition, 49*, 211–233.

James, W. (1892/1930). *Psychology: Briefer Course*. New York: Henry Holt & Co.

Kamath, H. O. (1988). *Neuropsychologia, 28*, 27.

Kellogg, R. T., Newcombe, C., Kammer, D., & Schmidt, K. (1996). Attention in direct and indirect memory tasks with short- and long-term probes. *American Journal of Psychology, 109*, 205–217.

LaBerge, D. (1995). *Attentional Processing: The Brain's Art of Mindfulness*. Cambridge, MA: Harvard University Press.

Mattingley, J. B., Davis, G., & Driver, J. (1997). Preattentive filling-in of visual surfaces in parietal extinction. *Science, 275*, 671–674.

McConkie, G. W., & Rayner, K. (1975). The span of the effective stimulus during a fixation in reading. *Perception and Psychophysics, 17*, 578–586.

McConkie, G. W., & Rayner, K. (1976). Asymmetry of the perceptual span in reading. *Bulletin of the Psychonomic Society, 8*, 365–368.

McConkie, G. W., & Zola, D. (1979). Is visual information integrated across successive fixations in reading? *Perception and Psychophysics, 25*, 221–224.

Moscovitch, M. (1995). Recovered consciousness: A hypothesis concerning modularity and episodic memory. *Journal of Clinical Experimental Neuropsychology, 17*, 276–290.

Mulligan, N. W., & Hartman, M. (1996). Divided attention and indirect memory tests. *Memory and Cognition, 24*, 453–465.

Newman, J. (1995). Thalamic contributions to attention and consciousness. *Consciousness and Cognition, 4*, 172–193.

Newman, J. (1997a). Putting the puzzle together part I: Towards a general theory of the neural correlates of consciousness. *Journal of Consciousness Studies, 4*, 47–66.

Newman, J. (1997b). Putting the puzzle together part II: Towards a general theory of the neural correlates of consciousness. *Journal of Consciousness Studies, 4*, 100–121.

Osaka, N., & Oda, K. (1991). Effective visual field size necessary for vertical reading during Japanese text processing. *Bulletin of the Psychonomic Society, 29*, 345–347.

Pizzamiglio, L., Guariglia, C., Incoccia, C., & Antonucci, G. (1990). Effect of optokinetic stimulation in patients with visual neglect. *Cortex, 26*, 535–540.

Pollatsek, A., Bolozky, S., Well, A. D., & Rayner, K. (1981). Asymmetries in the perceptual span for Israeli readers. *Brain and Language, 14*, 174–180.

Posner, M. I. (1994). Attention: The mechanisms of consciousness. *Proceedings of the National Academy of Science USA, 91*, 7398–7403.

Posner, M. I., Sandson, J., Dhawan, M., & Shulman, G. L. (1989). Is word recognition automatic? A cognitive-anatomical approach. *Journal of Cognitive Neuroscience, 1*, 50–60.

Ramachandran, V. S., & Cobb, S. (1995). Visual attention modulates metacontrast masking. *Nature, 373*, 66–68.

Rapcsak, S. Z., Watson, R. T., & Heilman, K. M. *Journal of Neurological Surgery, 50*, 1117.

Raymond, J. E., Shapiro, K. L., & Arnell, K. M. (1992). Temporary suppression of visual processing in an RSVP task: An attentional blink? *Journal of Experimental Psychology: Human Perception and Performance, 18*, 849–860.

Rayner, K., & Pollatsek, A. (1987). Eye movements in reading: A tutorial review. *Attention and Performance XII* (pp. 327–362). London: Erlbaum.

Rayner, K., Well, A. D., & Pollatsek, A. (1980). Asymmetry of the effective visual field in reading. *Perception and Psychophysics, 27*, 537–544.

Richardson-Klavehn, A., Gardiner, J. M., & Java, R. I. (1994). Involuntary conscious memory and the method of opposition. *Memory, 2*, 129.

Robertson, I. (1989). Anomalies in the laterality of omissions in unilateral left visual neglect: Implications for an attentional theory of neglect. *Neuropsychologia, 27*, 157–165.

Robinson, D. L., Bowman, E. M., & Kertzman, C. (1995). Covert orienting of attention in Macaques: II. Contributions of parietal cortex. *Journal of Neurophysiology, 74*, 698–712.

Rubens, A. (1985). Caloric stimulation and unilateral neglect. *Neurology, 35*, 1019–1024.

Saarinen, J. (1993). Focal attention and the perception of pattern structure in extrafoveal vision. *Scandinavian Journal of Psychology, 34*, 129–134.

Shulman, G. L., & Wilson, J. (1987). Spatial frequency and selective attention to local and global information. *Perception, 16*, 89–101.

Sperling, G. (1960). The information available in brief visual presentation. *Psychological Monographs, 74* (11).

Strehler, B. L. (1991). Where is the self? A neuroanatomical theory of consciousness. *Synapse, 7*, 44–91.

Szymanski, K. F., & MacLeod, C. M. (1996). Manipulation of attention at study affects an explicit but not an implicit test of memory. *Conscious and Cognition, 5*, 165–175.

Tsal, Y. (1994). Effects of attention on perception of features and figural organization. *Perception, 23*, 441–452.

Tsal, Y., Nachson, M., & Lamy, D. (1995). Towards a resolution theory of visual attention. *Visual Cognition, 2*, 313–330.

Underwood, N. R., & McConkie, G. W. (1985). Perceptual span for letter distinction during reading. *Reading Research Quarterly, 20*, 153–160.

Walker, R., Findlay, J. M., Young, A. W., & Welch, J. (1991). 'Disentangling neglect and hemianopia. *Neuropsychologia, 29*, 1019–1027.

Ward, R., & Goodrich, S. (1996). Differences between objects and nonobjects in visual extinction: A competition for attention. *Psychological Science, 7*, 177–180.

Wong, E., & Weisstein, N. (1983). Sharp targets are detected better against a figure, and blurred target are detected better against a background. *Journal of Experimental Psychology: Human Perception and Performance, 9*, 194–202.

Part III

Neural philosophy

Chapter 8

Recent work on consciousness

Philosophical, theoretical, and empirical

Paul M. Churchland and Patricia S. Churchland
University of California, San Diego

The loyal opposition

The possibility, or impossibility, of a neurobiological account of consciousness has been a hot topic of late, in both the public and the academic presses. A variety of theoretical overtures on the positive side (Baars 1988; Churchland 1995; Crick & Koch 1990; Dennett 1991; Edelman 1989; Llinas et al. 1994; Llinas & Ribary 1993) have provoked a gathering resistance focused on a single topic, namely, the ineffable qualitative characters of one's sensations, and their accessibility to no one but oneself. To use the current code words, the core problem for all aspiring physical accounts is said to be the undeniable existence of sensory qualia and the fact of their unknowability by any means except the subjective, introspective, or first-person point of view. This is sometimes referred to as "the hard problem", to distinguish it from the allegedly easier problems of memory, learning, attention, and so forth.

Sensory qualia are thought by many (Chalmers 1996; Jackson 1982; McGinn 1982; Nagel 1974; Searle 1992) to pose a problem for physicalist accounts – indeed, an insuperable problem – as follows. The intrinsic qualitative character or quale (pronounced kwah-lay) of a pain is to be sharply distinguished from the many causal, functional, and relational features of a pain. Examples of the latter are familiar: pain is typically caused by bodily stress or damage; pain typically causes unhappiness and avoidance behavior; pain disappears under analgesics. These are all extrinsic features of pain, accessible to everyone. They can be revealed in public experiments; they are known to every normally socialized person; and they play a central role in our public explanations and predictions of one another's behavior. Moreover, these causal/relational features of pain form a legitimate target for the reductive/explanatory aspirations of a growing neuroscience. The only requirement is to discover whatever states of the brain display exactly the same causal/relational profile antecedently accepted by us as characteristic of the state of pain. This is what would constitute, here, as elsewhere in science, a reductive explanatory account. If the discovered parallel of profiles were suitably systematic, we could then legitimately claim to have discovered what pains really are: they are states of the brain, the states captured by the neurobiological theory imagined.

Or rather, we could make that claim, except for one glaring difficulty: the intrinsic quale of pains would inevitably be left out of such an account, as would each person's unique knowledge of his pain qualia. It is evident, runs the argument, that our reductive aspirations here confront a problem not encountered in any other scientific domain. Specifically, the phenomena involved in conscious awareness are not exhausted by the fabric of causal and other relations in which they are embedded. In all other domains, the essential nature of the phenomena are so exhausted. But in the unique case of conscious awareness, there is a family of intrinsic properties – the subjective qualia that enliven our inner lives – whose essence cannot be captured in any causal, functional, structural, or relational story.

To drive its point home, the antireductionist argument continues as follows. The sciences, or at any rate, the physical sciences, are limited by nature to providing stories of this latter kind, stories that reconstruct the causal/relational reality of the target phenomena. They are therefore doomed to failure in precisely the case at issue, for the subjective qualia of one's sensations are something distinct from and additional to whatever causal/relational roles those sensations might happen to play in one's overall biological and cognitive economies. Such intrinsic qualia, though easily discriminable from the first-person point of view, are structureless simples in and of themselves. Accordingly, they can offer no "reconstructive purchase" for the aspirations of any physical science. They are the wrong kind of explanatory target. They are (i) metaphysically simple and (ii) exclusively subjective, whereas any physical reconstruction of them would have to be (i) based on causal/relational structures and (ii) entirely objective.

Thus the core resistance of the loyal philosophical opposition. A survey of the literature will find a variety of "thought experiments" whose function it is to illustrate or highlight, in some way or other, the essentially elusive nature of our subjective qualia, as they are addressed by any third-person, physicalist point of view. For example, we have Nagel's impenetrably alien bat (Nagel

1974), Jackson's neurobiologically omniscient but color-blind Mary (Jackson 1982), Chalmer's functionally normal but qualia-deprived zombies (Chalmers 1996), and everyone's puzzle about whether one's subjective color space might be inverted relative to that of one's fellows. Each such story depends, in its own way, on the set of convictions outlined in the preceding two paragraphs. In the end, it is these shared convictions that do the real anti-reductive work. Let us take a closer look at them.

The loyal opposition disarmed

We must concede, I think, that the quale of a pain, or of the sensation of red, or of the taste of butter, certainly seems to be simple and without hidden structure. So far as my native introspective capacities are concerned, they reveal nothing whatever in the way of constituting elements or relational structure. I can recognize and discriminate such qualia – spontaneously, reliably, and noninferentially – but I am quite unable to say how or on what basis I am able to identify them. So far, then, they seem to be intrinsic simples. And my knowledge of them would seem to be as direct and as fundamental as it could possibly be.

But we are here in danger of being seduced by that favorite of freshman logic classes: an Argument from Ignorance. That we are unaware of any hidden structure in the qualia of our sensations, that we do not know how our conscious awareness manages to discriminate among different sensations, these admitted facts about our own lack of knowledge do not entail that such qualia must therefore be free of underlying structure, nor that our first-person recognition of such qualia does not depend on some mechanism keyed to the relational features of that underlying structure.

Indeed, they do not even suggest such a view, because we know by sheerly logical considerations that, for any cognitive creature capable of discriminating among aspects of its experience, there must be some relatively low level of property-discrimination that is currently "basic" for that creature, in the weak sense that he is ignorant of and unable to articulate the possibly quite complex underlying causal basis on which those discriminations are made. The denial of this inevitable limitation would commit us to an infinite regress of discriminative levels, for each of which levels the creature can always articulate an account of the yet more basic level of properties on which his discriminations within the target level of properties consciously depends. Barring such infinities, for each creature there must be some level of property discrimination where his capacity for articulation runs out.

This logical fact is reflected in many familiar limitations. For example, you can recognize your youngest child's voice. You can discriminate it instantly from hundreds of her schoolchums. But you cannot articulate the acoustic basis of that discrimination. You can recognize suspicion in a person's face, instantly. But you will do very poorly at articulating the complex facial profile that prompts that judgement. Such judgements plainly have an underlying physical basis, but our ability to make them does not require that we be consciously aware of that basis.

It is both familiar and inevitable, then, that we be ignorant, at the level of conscious awareness, of the basis or etiology of at least some of our property-discriminations. And those inarticulated properties may well present themselves to us as being "simple", however complex their underlying nature might be. And this is so for reasons that have nothing to do with any issues that might divide materialists from dualists. We should not be too quickly impressed then, that our discrimination of the qualia of our sensations turns up as constituting such a currently "basic" level. The "simplicity" of our sensory qualia need reflect nothing more than our own predictable ignorance about how we manage to discriminate them.

This does not show that the antireductionists must be wrong. But it does show that one of their crucial premises – concerning the intrinsic, nonrelational, metaphysical simplicity of our sensory qualia – is a question-begging and unsupported assumption. Whether qualia have that character is a matter for ongoing research to discover, one way or the other, not a matter on which partisan enthusiasts can pronounce from the armchair.

There is a further reason we should resist the premise of structureless simplicity. There is a robust historical precedent for the scientific issue here at stake. It is instructive to recall it.

Consider a familiar family of qualitative features, not of our internal states, but of external physical objects. Specifically, consider the colors that enliven the surfaces of apples and flowers and stones. These, too, are features we can learn to discriminate – spontaneously, reliably, and noninferentially. These, too, are features that present no internal causal/relational structure to our conscious apprehension. These, too, are features that are discriminationally "basic" for us in the weak sense outlined above. These, too, have been described as something we apprehend "directly" (the position is called "Direct Realism," and it survived into the present century). And these, too, have been defended as physically irreducible "simples" (by thinkers such as the English poet, William

Blake, and the German naturalist, Wolfgang Goethe, both of whom were reacting against Sir Isaac Newton's newly published particle theory of light). Finally, these features, too, might have been plausibly claimed to be epistemically accessible from but a single epistemic standpoint, viz., the "visual point of view." (Call such epistemically-isolated properties "visjective" properties.)

Despite all these antireductive presumptions, a century's scientific research has taught us that the surface colors of external objects do indeed have a rich internal causal/relational structure, a structure, moreover, that is systematically related to our native mechanisms for color discrimination. External color qualia, at least, are not metaphysical simples at all, despite a fairly convincing first impression. To a first approximation, an object's color is constituted by its differential capacity to absorb and reflect diverse wavelengths of incident electromagnetic radiation. Colors have been successfully reduced to, or better, reconstructed as, appropriate families of reflectance vectors. The "trichromatic" retinal cells of humans are differentially responsive to these reflectivity profiles of external objects, and that is how we discriminate objective colors.

The lessons here are obvious, but worth enunciating. While it may occasionally seem plausible to insist on a heavy-duty metaphysical distinction between the "causal/relational" and the "intrinsic" features of some domain of phenomena, that distinction may well be ignorance-driven and entirely empty. And while it may seem doubly plausible to insist on some such distinction when we also possess a native epistemic access to precisely the "intrinsic" features of the relevant domain, that native access does nothing whatever to assure the problematic contrast at issue. It need mark nothing but the current limits of our discriminatory understanding.

In sum, nothing about the case of our internal sensory qualia guarantees their structureless simplicity, nor even suggests it. In fact, such historical precedents as lie at hand (see above) suggest precisely the opposite conclusion. Our subjective sensory qualia may each have a rich internal structure, and thus it remains entirely possible that a matured neuroscience eventually discover that hidden structure and make it an integrated part of the overall reconstruction of mental phenomena in neurobiological terms. On this, more below.

But what of the peculiarly exclusive character of one's epistemic access to sensory qualia? What of their inaccessibility to any apprehension or instrumental detection from the objective or "third-person" point of view? Does not this show, all by itself, that internal qualia must lie beyond the reach of any physicalist reduction?

No, because this exclusivity premise is no more substantial or compelling than the simplicity premise we have just dismissed. To illustrate, recall once more the case of the external colors and our "exclusively" visual access to them. The theoretically motivated claim that external colors might also be detected by some nonvisual artificial instruments – instruments that respond to an object's electromagnetic reflectance vector – would no doubt be dismissed by Blake and Goethe as a case of confusing the essentially "visjective" property of color with the wholly "nonvisual" property of electromagnetic reflectance profiles. Alternatively, our two antireductive stalwarts might simply dismiss such a claim as "changing the subject" away from the properties that really interest us (the visible colors) and toward something else (reflectivity profiles) that is only contingently related to colors, at best. Such digging in of the heels is perhaps only to be expected. So long as one is antecedently convinced that external colors are ontologically distinct and physically irreducible properties, then so long will one be inclined to dismiss any novel and unexpected epistemic access to them as being, in truth, an access to something else entirely.

But it needn't be so. In the case of external colors, it is now plain that it wasn't so. Insofar as color exists at all, to access an object's reflectivity profile is to access its objective color. And in the currently spotlighted case of internal sensory qualia, we must be sensible that something similar may happen. When the hidden neurophysiological structure of qualia (if there is any) gets revealed by unfolding research, then we will automatically gain a new epistemic access to qualia, above and beyond each person's native and currently exclusive capacity for internal discrimination. As aspiring reductionists, we cannot insist that this will happen. That is for research to determine. But neither can the antireductionist insist that it will not. His second premise, therefore – the epistemic exclusivity premise – is as question-begging and as ignorance-driven as the first.

We conclude that no substantive considerations currently block the reductive and explanatory aspirations of the neurosciences. The celebrated arguments that attempt to do so derive their plausibility from our current ignorance, from our limited imaginations, and from a careless tendency to beg the very questions at issue. We should not be taken in by them.

A final caution. A complete knowledge of the neurophysiological basis of our conscious awareness of sensory qualia – such as might be acquired by Frank Jackson's utopian neuroscientist, Mary (Jackson 1982) – would of course not *constitute* our normal native capacity for the spontaneous, noninferential discrimination and recognition of sensory qualia. Though aimed at the same epistemic target – sensory qualia – this native capacity is a distinct cognitive capacity. (Evolutionarily, it antedates our capacity for discursive science by hundreds of millions of years.) Mary could have the scientific story down pat, and yet, because of a congenital color-blindness perhaps, lack the native capacity entirely. This possible dissociation has often been appealed to as proving that phenomenal facts about consciousness are distinct from physical facts about the brain (see again both Jackson 1982 and Chalmers 1996). But it shows nothing of the sort. The dissociation described is entirely possible – indeed, it is cleanly explicable – even on a purely physical account of mental phenomena. It can hardly entail, therefore, the falsity of reductive materialism.

Phenomenal qualia: Theory and data

Defending the abstract possibility of a neurobiological reduction of mental phenomena is one thing. Actually providing such an explanatory reduction is quite another. These closing sections address work that is already underway. We shall try to sample some recent results as they bear on the antireductive worries so widespread in the literature.

We begin with what seems to us an irony: of the many problems confronting our explanatory ambitions in neuroscience, the so-called "hard problem" – the problem of subjective sensory qualia – appears to be one of the easiest and closest to a stable solution. As we see it, the much harder family of problems is the clutch that includes short-term memory, fluid and directable attention, wake versus sleep, and the unity of consciousness. These come up in the next section. For now, let us examine some representative work that bears on qualia.

Neurobiology aside, psychophysics has long been in the business of exploring the structure of our perceptual "quality spaces". Without ever sinking an electrode, one can systematically explore the relative similarity relations and betweenness relations among hundreds of color samples, as they are spontaneously judged by a large sample of experimental subjects. For each subject, those judgements allow the experimenter to locate every color sample in a unique position relative to all of the others. For normal humans, the structure of that quality space is well-defined and uniform across subjects. It is roughly as portrayed in Figure 1. (A similar story can told for the other sensory modalities, but we here choose to mine a single example in some depth.)

Here we have, already, some systematic structure that unites the set of visual qualia, a structure that the neurobiology of vision can hope (and must try) to reconstruct. Incidentally, the "inverted spectrum" puzzle is here quickly characterized. Evidently, the mirror-image of the double cone in Figure 1 would capture the internal relations of that color space equally well, and one

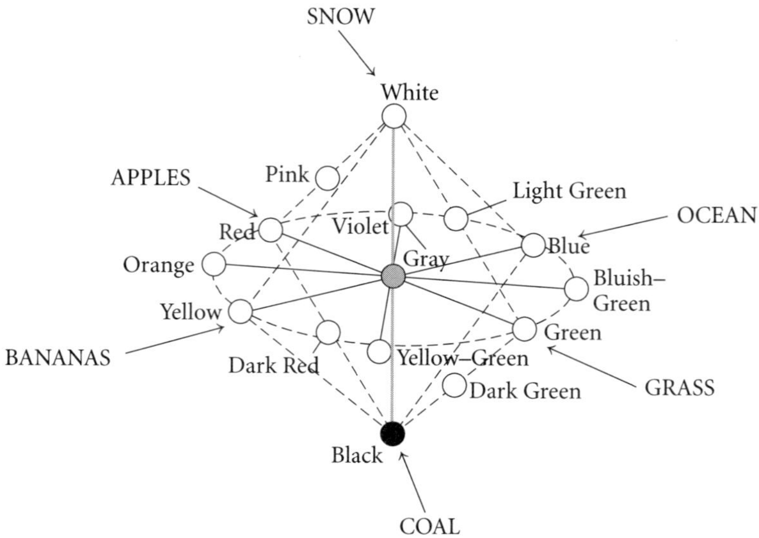

Figure 1. The intrinsic order of the human quality space for color. Some of the standard causal relations that connect it to the external world are indicated at its periphery. (A caution: this figure wrongly portrays our quality space as symmetric or homotopic. Importantly, it is not, and its internal asymmetries provide one way to address worries about so-called "inverted spectra." Hardin (1988) and Clark (1993) provide the best discussions on this topic. But I am here trying to pose the problem in its most difficult form, in order to outline an independent solution.)

might wonder whether, and how to tell whether, one's own color space is causally attached to the external world in the same way as everyone else's. Single-axis inversions are not the only possibility here: partial rotations of the entire space, around arbitrary axes, are quite conceivable also. One wonders how we can get an independent handle on everyone's internal color space in such a way as to settle such questions.

One possible way is to find some determinate and accessible internal or intra-qualia structure that is variable across qualia, a structure that gives rise to the same inter-qualia relations as are displayed in Figure 1. This internal structure would give us a means of identifying an individual internal quale, a means that is independent of both its relations to other internal qualia and its problematic causal connections with external stimuli. The existence of such an intra-qualitative structure, of course, will be rejected as impossible by the antireductionist of the preceding sections, but we have learned not to be buffaloed by such insistence. Let us now look at a live candidate.

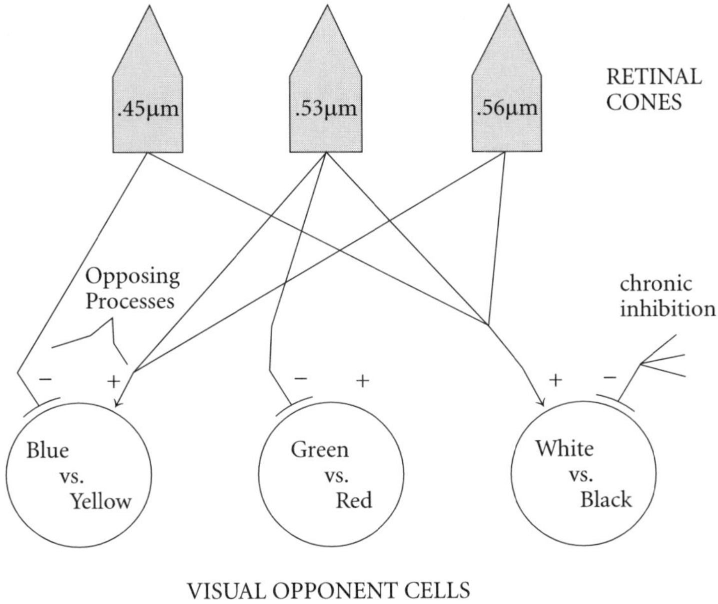

Figure 2. A schematic of the proposed visual opponent processes for coding color in the human and primate visual systems. (After Clark 1993; Hardin 1988; and Hurvich 1981.)

Our current "best theory" for the neuroanatomy and neurophysiology of human and primate color vision is the so-called opponent processes theory. On this proposal, color is coded as a three-element vector of activation levels within three proprietary types of cells synaptically downstream (perhaps in the LGN) from the familiar three types of retinal cone cells. Those three downstream cell-types are each the locus of an excitatory/inhibitory tug-of-war (hence, "opponent processes"), as indicated in Figure 2. As a result of the connections there portrayed, the left-most cell type ends up coding the relative dominance of yellow light over blue; the middle cell type ends up coding the relative dominance of red light over green; and the right-most cell type ends up coding the relative dominance of brightness over darkness roughly summed over all wavelengths.

Here we can determine, experimentally, exactly what signature activation triplet is produced across the opponent-color cells by any given external color sample. We have, that is, an independent access to the internal structure of the creature's neurophysiological response, and we can make determinate judgements of what external objects produce that response. Let us now ask how those

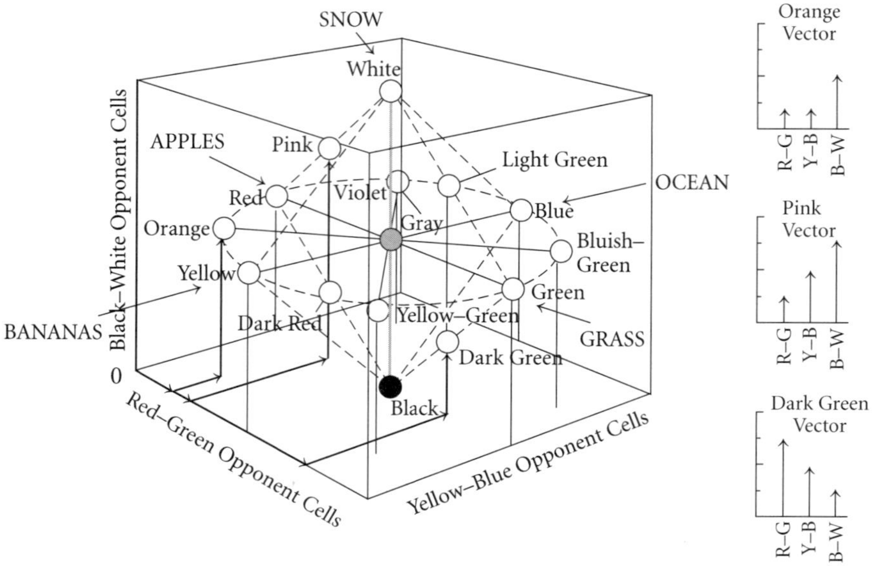

Figure 3. The vector space of opponent-cell coding, with some of the standard causal connections it bears to the external world. Note the isomorphism with the quality space of Figure 1, both in its internal relations and its external connections. Sample vectors – for orange, pink, and dark green – are displayed as histograms at the side. These vectors are also traced within the color space proper.

proprietary coding triplets are globally organized, within the vector space of all possible triplets, according to their Euclidian proximity to and distance from each other. The answer is portrayed in Figure 3, and what is salient is that the coding scheme at issue produces the same global organization of features found in the psychophysically determined quality space of Figure 1, an organization that enjoys the same pattern of causal connections to the external world.

As in the historical case of external color qualities, such a systematic coincidence of causal and relational structures suggests a reductive explanation of the original quality space, and a corresponding identity claim, viz., that one's internal or subjective color qualia are identical with one's coding triplets across the three types of opponent cells. If that is what sensory qualia really are, then we have, in the story just outlined, a systematic explanation of the specific contents and rich internal structure of our native quality space for color, an explanation that also accounts (via the electromagnetic theory of light) for its specific causal connections with the objective colors of objects in the external world.

Does this explanation allow for the possibility of "inverted spectra", as mooted in the traditional philosophical puzzle? Indeed it does, and it even entails how to produce that effect. Take a normal adult observer and, while holding all other neural connections constant, take all of the axons currently synapsing onto his blue/yellow opponent cells and exchange them with the axons currently synapsing onto his red/green opponent cells. That will do it. Upon awakening from this (highly fanciful) operation, our observer will see things quite differently, because his internal color space has been surgically remapped onto the external world. It has been "rotated" ninety degrees around the black/white axis, relative to the "external causes" in Figure 1.

Alternatively, a mirror inversion of the original space, along one axis, could be achieved by an (even more demanding) surgery that inverted the polarity of all of the existing synapses onto exactly one of the three populations of opponent cells. Of course such "inverted spectra" are conceptually possible! It is even possible that we might detect them experimentally in the details of our neural wiring, or manipulate them as indicated .

Notice that the present account also entails the existence of a variety of suboptimal color spaces for that small percentage of humans who are missing one or other of the normal populations of retinal cones. Their opponent cells thus end up denied important dimensions of information. If this vector-coding approach to qualia is correct, then psychophysical mapping of these nonstandard quality spaces will also have to match up with these nonstandard vector spaces at the cellular level. So far as I am aware, such experimental work on the varieties of color-blindness remains to done, at least as concerns the cellular level, but it offers a clear test of the theory.

The vector-coding approach to the inner nature of sensory qualia can be further tested as follows. As mentioned earlier in a figure legend, the normal human quality space for color is importantly asymmetric. The actual profile of that asymmetry is something that the opponent-cell theory must also aspire to explain, perhaps in terms of the nonuniform response profiles of the human opponent cells involved.

Should our growing neurophysiological reconstruction of human color space succeed in accounting even for this level of detail, then we would have a gathering case that our internal color qualia are indeed identical with coding vectors across the opponent cells in our LGN. It would not cinch the case, but we would here have the same sorts of systematic evidence that justify reductive explanatory claims elsewhere in science. And, dualist enthusiasms aside, it should be accorded the same weight that such evidence possesses elsewhere in science. No more. But no less.

Another caution is in order here. For we are not urging the truth of the preceding account as the refutation of the antireductionist's trenchant skepticism. Sensory qualia may yet be metaphysical simples forever inaccessible to neuroscience. But the antireductionist arguments addressed at the beginning of this paper made the very strong claim that (1) we already know that they are structureless simples, and that (2) no theory from the physical sciences has any hope of even addressing the phenomena in this area. Against this, we are claiming only that (1) we know no such thing, and (2) a perfectly respectable theory from the physical sciences already addresses the relevant phenomena, in considerable detail and with some nontrivial success.

Our caution is well-advised for a further reason. While the opponent-cell story appears correct as far as it goes, it may be that we should resist identifying our conscious color quality space with the specific vector space of those relatively peripheral LGN opponent cells. They may be entirely too peripheral. A more plausible candidate vector space, for the physical embodiment of our color qualia space, might well be the activation space of some neuronal population two or three synaptic steps downstream from the LGN's opponent cells, a subpopulation of the primary visual cortex, for example, or perhaps the cells of area V4. Such higher-level vector spaces, boasting more highly-processed information, may allow a more faithful reconstruction of the fine-grained features of our actual color qualia space, both in normals and in abnormals. And they may locate the relevant vectors in a processing arena more plausibly implicated in the phenomenon of conscious awareness. To which topic we now turn.

Consciousness and attention: Theory and data

Sensory qualia aside, why and how does anything at all ever make its way into conscious awareness? Vector coding is something that occurs at all levels of the nervous system, in thousands of distinct neuronal populations which represent an enormous range of information concerning the state of one's body and the state of the world around it, only a small percentage of which information is ever present to conscious awareness. What distinguishes this preferred class of conscious representations from the much larger class of representations that never see the light of consciousness?

This matter is no longer the smooth-walled mystery it used to be. A growing appreciation of the brain's neuronal organization, and enhanced access to brain activity in awake behaving creatures, has produced an environment

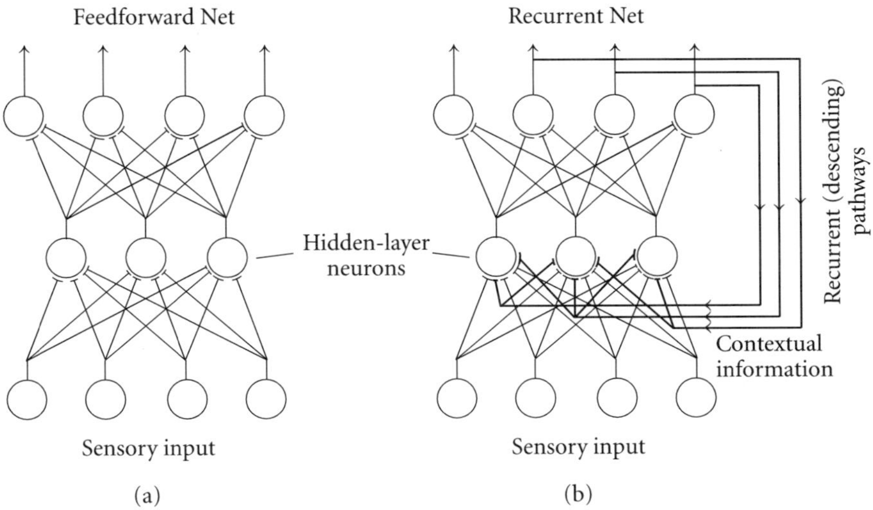

Figure 4. (a) The architecture of a standard feedforward network. Its response, at the middle layer, to a specific input activation vector, is some unchanging activation vector across that second population. (b) The architecture of a recurrent network. Its response, at the middle layer, can be an unfolding sequence of activation vectors, since that second population is also subject to subsequent and repeated input activity returned to it from the downstream cell population.

in which responsible theoretical activity can at least get off the ground. The critical elements, it seems to us, are as follows.

Since the mid-80s, there has been a groundswell in the construction of artificial neural networks and in the exploration of their pattern-recognizing capacities. A prototypical feedforward architecture appears in Figure 4a. Modelers quickly learned, however, that for the great many important patterns that have a temporal dimension – such as typical motor sequences and typical causal processes – a purely feedforward network is representationally inadequate. One must move to networks that have descending or recurrent pathways, as in Figure 1b, in addition to the standard ascending ones.

These additional pathways yield a network whose vectorial response, at the middle layer, to sensory-layer input, is partly a function of the concurrent activational or cognitive state of the third layer, which state is the result of still earlier inputs and earlier processing. A recurrent network's response to a given stimulus, therefore, is not fixed solely by the structural features of the network, as in a feedforward network. The response varies as a function of the prior dynamical or cognitive context in which the input stimulus happens to occur.

This middle-layer response can also unfold continuously over time, as the middle-layer cells receive an ever-changing set of modulating influences, from the downstream cognitive activity, via the recurrent pathways. For such a recurrent system, the middle-layer response is typically a sequence of activation vectors. That is to say, the response is a trajectory in activation space, rather than a point.

What is charming about such networks is that, by adjusting the weights and polarities of their synaptic connections appropriately, we can train them to respond to various input stimuli with proprietary activational trajectories. These vector sequences can represent salient causal sequences in the network's perceptual environment, sequences that the network has been trained to recognize. Alternatively, they can serve as generators for well-honed motor sequences that the network has been trained to produce. In both the perceptual and the motor domains, therefore, recurrent networks achieve a command of temporal patterns as well as spatial patterns – a command of time as well as space. More generally, they hold out a powerful set of theoretical resources with which to understand the phenomena of learning, memory, perception, and motor control.

They may also help us get an opening grip on consciousness, as follows. If a reductive explanatory account of consciousness is our aim, then here, as elsewhere in science, we must try to reconstruct the known features of the target phenomena using the resources of the more basic science at hand. (See Churchland & Churchland 1990 for an accessible account of the nature of intertheoretic reduction.) What features of consciousness stand out as salient targets for such reconstruction? While they are hardly exhaustive, the following would likely be included on anyone's list.

Consciousness involves short-term memory. Consciousness does not require concurrent sensory input. Consciousness involves directable attention. Consciousness can place different interpretations on the same sensory input. Consciousness disappears in deep sleep. Consciousness reappears, in somewhat altered form, in dreaming. Consciousness brings diverse sensory modalities together in a single, unified experience.

Recurrent networks have reconstructive resources that might eventually prove relevant to all of the elements of this list. First, any recurrent net already embodies a form of short-term memory: the information embodied in the middle-layer activation vector gets processed at the next layer and is then sent back to its origin, perhaps in modified form. That information may cycle through that recurrent loop many times while it slowly degrades. If some of it is important, the network may even be configured to retain that information, undegraded, through many cycles, until the unfolding cognitive context no longer values it. Such a system provides, automatically, a short-term, content-sensitive, variable decay-time memory.

A recurrent network can also engage in cognitive activity in the absence of current stimulation at the input layer, because activity in the recurrent pathways can be sufficient to keep the system humming away all on its own. It can also modulate, via those same pathways, the manner in which it reacts to sensory-layer stimuli, and the salience that certain aspects of that input will command. This gives us a crude analog for both the plasticity of conceptual interpretation, and the capacity for steering one's perceptual attention.

Moreover, a recurrent network can have its recurrent pathways selectively disabled for a time, and it will temporarily revert to a purely feedforward network, in which condition it will lose or suspend all four of the cognitive capacities just described. Perhaps deep sleep in humans is some peculiar instance of this highly special form of cognitive shutdown. Correlatively, it may well be that dreaming consists in the spontaneous or self-driven activity of a richly recurrent network that wanders among the learned trajectories that have come to dominate its normal waking operation, a meander that is temporarily free of the coherent guidance of sensory input from a stable external world, a meander that is also temporarily detached from the motor effectors it would normally drive.

Finally, a recurrent network can integrate information from different sensory modalities by delivering such information, directly or indirectly, back to a common cell population. The activations vectors at such a population will thus embody multimodal information. The Damasios (Damasio 1995) call such brain areas "convergent zones."

Though vague, these assembled reconstructive suggestions are more than just suggestions. We know that the brain is a profoundly recurrent neuronal system, and there is experimental evidence that provides some initial support for the conjectures about the dynamical contrasts between deep sleep, dream sleep, and the wake state (Llinas & Ribary 1993; Llinas et al. 1994). Guided by neuroanatomy, we can at least see how we might sharpen the vague proposals just displayed. And guided by neurophysiology, we can at least see how progressively to test them.

A possible answer, therefore, to the question that opened this section, is as follows. A representation is an element of one's current conscious awareness just in case that activation vector occurs at the focal population of a suitably central recurrent system in the brain, a system that unites the several sensory modalities and dominates the control of motor behavior.

This may or may not be true. But it is a reductive hypothesis that explicitly addresses both of the central concerns of the antireductionist, the concerns that introduced this paper. First, we can indeed give an illuminating physical account of the "intrinsic" nature of our various sensory qualia. In short, they are activation vectors, one and all. And second, we can indeed suggest a possible account of when such vectors are a part of one's current conscious awareness. They must occur as part of the representational activity of a suitably recurrent network meeting the functional and anatomical constraints outlined above. The antireductionist's counsel of despair was never well-founded, as we saw earlier in our second section. And it came with no competing framework with which to sustain a viable tradition of research. Best we should put that counsel behind us then, and go where the recent progress pulls us.

References

Baars, B. J. (1988). *A cognitive theory of consciousness*. Cambridge, UK: Cambridge University Press.

Chalmers, D. (1996), *The Conscious Mind*. Oxford, UK: Oxford University Press.

Churchland, P. M. (1995). *The engine of reason, the seat of the soul: a philosophical journey into the brain*. Cambridge, MA: MIT Press.

Churchland, P. M., & Churchland, P. S. (1990). Intertheoretic reduction: a neuroscientist's field guide. *Seminars in the Neurosciences, 2*, 249–256.

Clark, A. (1993). *Sensory qualities*. Oxford, UK: Clarendon Press.

Crick, F., & Koch, C. (1990). Towards a neurobiological theory of consciousness. *Seminars in the Neurosciences, 2*, 263–276.

Damasio, A. (1995). *Descartes' error*. New York, NY: Putnam and Sons.

Dennett, D. C. (1991). *Consciousness explained*. Boston, MA: Little Brown.

Edelman, G. E. (1989). *The remembered present: A biological theory of consciousness*. New York: Basic Books.

Hardin, C. L. (1988). *Color for philosophers: Unweaving the rainbow*. Indianapolis: Hackett.

Hurvich, L. M. (1981). *Color vision*. Sunderland, MA: Sinauer.

Jackson, F. (1982). Epiphenomenal qualia. *Philosophical Quarterly, 32*, 127–136.

Llinas, R., & Ribary, U. (1993). Coherent 40-Hz oscillation characterizes dream state in humans. *Proceedings of the National Academy of Sciences, USA, 90*, 2078–2081.

Llinas, R., Ribary, U., Joliot, M., & Wang, X. J. (1994). Content and context in temporal thalamocortical binding. In G. Buzsaki et al. (Eds.), *Temporal coding in the brain*. Berlin: Springer-Verlag.

McGinn, C. (1982). *The subjective view: Secondary qualities and indexical thoughts*. Oxford, UK: Oxford University Press.

Nagel, T. (1974). What is it like to be a bat? *Philosophical Review, 83*, 435–450.

Searle, J. (1992). *The rediscovery of the mind*. Cambridge, MA: MIT Press.

Part IV

Quantum mind

Chapter 9

Quantum processes in the brain*

A scientific basis of consciousness

Friedrich Beck and John C. Eccles
Darmstadt University of Technology, Germany

1. Introduction

Since the earliest testimony of mankind we can observe a deeply mysterious feeling about the invisible soul, as contrasted to the material body. Originally, the soul was conceived as material, as air, or as finest matter penetrating the body. Later on, in the Greek philosophy of Plato and Aristotle the soul was conceived as a non-material entity that interacts with the body. This *dualistic* concept raised no problem with Greek natural philosophy which had rather abstract and primitive ideas about cause and effect. At the same time the first great physician, Hippocrates, stated that in movement the brain is the interpreter of consciousness, it tells the limbs how to act, and it is also the messenger to consciousness, expressing *dualism* and *interactionism*.

There was not much change to these ideas until philosophy turned to rationalism at the end of the Renaissance period. It was the great French philosopher and mathematician René Descartes (1596–1650) who presented his well-known explanation of mind-body interaction (Descartes 1644). '*cogito ergo sum*' was the shortest possible rational expression of mind-body entity, realized by interaction of a non-material mind, *res cogitans,* with a material brain, *res extensa.* Descartes unfortunately combined this abstract rationalism with the inadequate postulate that the pineal body was the organ immediately moved by push from the human soul. This brought the *Descartes' dualism* into heavy criticism, even among contemporaries like Leibniz and Spinoza.

Descartes' dualism cam into even heavier waters with the triumph of modern science. At the end of 19th century classical physics in the form of Newton's mechanics and Maxwell's electrodynamics was regarded as a complete, closed and causal description of the world, leaving no room for any kind of freedom. This is best expressed in the form of *Laplace's daemon*: Giving a super-mind the momentary initial conditions of the whole world, he can calculate all its future unambiguously. The world runs like a clock! There was then not much room left for a dualistic picture of mind-brain interaction, and materialism prevailed as has been expressed rather drastically by Charles Darwin: "Why is thought being a secretion of the brain, more wonderful than gravity a property of matter?" (Gruber 1974).

Materialism was, however, not accepted by everybody. Too strong was the believe, based on personal experience, that *self-consciousness* governs our actions in the world, and that this requires the ability for free, not pre-determined, decisions. Natural scientists were quite aware that such a non-materialistic view caused an unsurpassable conflict with the laws of nature to which our bodies, including the brain, underlie as material biological objects. The frustration could not be better expressed than in an address of the neurophysiologist and science philosopher Emil Du Bois-Reymond which he presented in 1872 at the German natural scientists and physicians meeting:

> There occurs at a certain point of evolution of life in the world, which we do not know and whose determination is of no importance in this connection, something new and hitherto incommensurable. Something which is, like the nature of matter and force and like the first motion, mysterious (...) This new mystery is consciousness. I shall now, as I believe in an unambiguous manner, outline that not only by our present day knowledge consciousness can not be explained out of its material conditions, what apparently everybody would admit, but that by its own nature, it will be never explainable from these conditions. (Du Bois-Reymond 1916, transl. by author.)

He ended his talk with the apodictic prognosis "*ignorabimus*".

The '*mystery*', as stated by Du Bois-Reymond, has survived up to within our days. It has been rationalized by the Austrian-British philosopher Karl Popper (1902–1994) in his 'three-world-classification' comprising all existents and experiences that has been achieved by mankind (Popper 1972, see Figure 1). There has been quite some misunderstanding of Popper's three worlds, insofar as they were regarded as *physically* separated, instead of *categorically* separated. This misunderstanding produced severe criticism of Popper's classification in tying him up with a primitive version of Cartesian dualism. Popper, however, was very well aware of the epistemological problems in mind-brain dualism, and shortly before his death he took the discussion up again (Popper et al. 1993; Lindahl & Århem 1994; Beck 1996a). A critical evaluation of the different posi-

World 1	**World 2**	**World 3**
Physical objects and states	*States of consciousness*	Objective knowledge
1. Inorganic: *Matter, energy*	Subjective knowledge	Cultural heritage:
2. Biology: *Structure and actions of all living beings, human brains*	*Experience of* *perception* *thinking* *emotions* *intentions* *memories* *dreams* *creative acts*	*philosophical* *theological* *scientific* *historical* *literary* *artistic* *technological*
3. Artefacts: *Material substrates of human creativity, tools, books, works of art*		Theoretical systems: *scientific problems* *critical arguments*

Figure 1. Tabular representation of the three worlds of Karl Popper, comprising the real world and the world of our experience.

tions, as presented in the contemporary debate on the nature of consciousness can be found in Eccles (1994).

An important qualitatively new aspect has been brought into the debate on the mind-brain problem when several authors realized that quantum physics frees physical processes from the strict determinism of the classical mechanistic picture, which has led Du Bois-Reymond to his *mystery*. The quantum aspect, however, opened only rather late a new pathway to understand consciousness, pioneered by Wigner 1964, and later followed on by several authors (Margenau 1984; Squires 1988; Eccles 1990; Donald 1990; Stapp 1991). Most influential in broader public discussions were the two books of Penrose (1984, 1994).

It should be emphasized, however, that such a discussion has two aspects which should be clearly separated. One resides on the epistemological level of quantum logic, which, in turn, is related to the interpretation of quantum mechanics, or even to the need for an essential extension of the present theory (Penrose 1994). The other aspect is the search for a better understanding of synaptic and neural actions, and their relation to microscopic, and eventually large scale coherent action, where quantum processes might work in a decisive manner on the basis of present-day theory. This is certainly open to experimental and theoretical research in contemporary brain physiology, and it is completely free from epistemological interpretations. Only very few realistics attempts have been made, however, to contribute to the latter question, and to locate a quantal process in the functional microsites of the neocortex (Beck & Eccles 1992; Hameroff & Penrose 1995).

In this article we discuss briefly the role of quantum logic, as contrasted to the deterministic logic of classical physics, and, in the main part, give a résumé of our work on the microsite hypothesis, and its relation to synaptic emission. We then introduce the quantum trigger model based on electron transfer in the biological reaction center of the synaptic membrane. Finally we discuss consequences for regulating the coherence patterns of groups of neurons, as has been observed in the visual cortex, or in positron emission tomography (PET) studies (Singer 1990; Pardo & Raichle 1991).

2. The epistemological question: Why are quantum processes interesting?

The dualistic concept of mind-brain interaction, even in its modern version as, e.g., presented by Popper (1972), involves the assumption that the immaterial mind acts upon the material brain. This would either require some kind of a force (Popper et al. 1993; Lindahl & Århem 1994; Beck 1996a), and this implies the mind to be not really immaterial, or one is forced to give up at least one of the global conservation laws of physics, based on strict space-time symmetries. The latter is not acceptable from a purely scientific standpoint. It should, however, be emphasized that the apparent contradiction between *dualism* and *identity theory* ("the mind is the brain") is itself deeply routed in the logic of a world view based on classical physics. We are so used to our daily experience of macroscopic events surrounding us, and which are well described by classical physics, that we can hardly appreciate a world in which determinism is not the rule of the game. *'If – then'* lies deeply in our conscious experience. Yet the quantum world, governing microphysical events, has a different logic.

The essence of quantum mechanics can be demonstrated in the very simple Young interference experiment, well known from wave optics (Figure2). Here, instead of optical waves, a beam of particles (e.g. electrons), described according to de Broglie's hypothesis (cf., e.g., Messiah 1961) by an incoming plane wave, hits a screen with two slits, which produces (Huygens' principle) two interfering waves with complex amplitudes A_1 and A_2. On the screen S they generate an interference pattern which results from the absolute square of the added amplitudes, $|A_1 + A_2|^2$. This is depicted by the piles behind the screen which indicate the number of particles registered by detectors at different positions. So far this experiment is completely analogous to the corre-

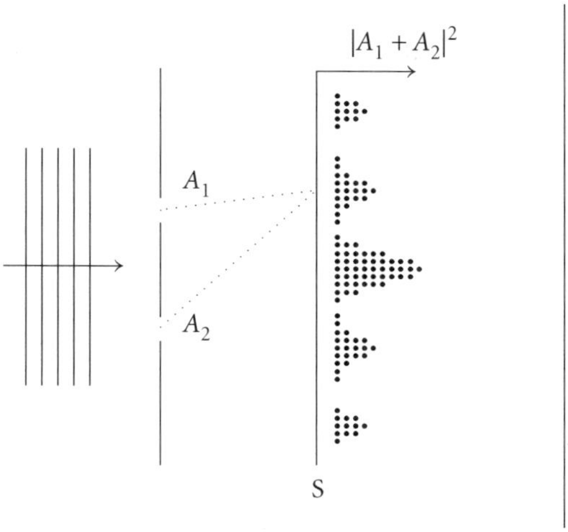

Figure 2. A particle wave incident on a wall with two openings. Behind the openings secondary waves are propagating with complex amplitudes A_1 and A_2. At the screen detectors measure for ensembles of many particles the intensity $|A_1 + A_2|^2$ which is schematically sketched. For a single event the outcome is undetermined. (The distance of the openings is of the same order as the de Broglie wavelength of the particles, after Fliessbach 1991.)

sponding experiment with light, where the outcome is the well known result of wave diffraction. Quantum mechanics, however, is a theory for the *single event*. What happens if one reduces the intensity of the particle beam, so that finally only one particle hits the screen S at a time? Now the occurrence of the particle at the screen is *completely undetermined*! This is the *non causal nature* of quantum mechanics for the outcome of a single event. It is the consequence of the '*von Neumann reduction of the wave function*', or state collapse which reduces the whole interference pattern to a single registration in one of the detectors (von Neumann 1955).[1] It comprises the *indeterminacy* of the future in the microscopic world. Quantum mechanics does this by subsequent interactions with the neighborhood,[2] and there is no need for an obscure 'orchestrated reduction' by background gravitational fields (Penrose 1994).

The basic difference between classical and quantum dynamics can be made clear on a somewhat more abstract basis in a simple diagram without entering into the formal subtleties of the theory. The generation of a physical process consists of preparing an *input* (the initial condition) followed by a more or less complicated *process*, leading to an *output* (the result) depending on initial conditions and the dynamics of the process. The output can be observed

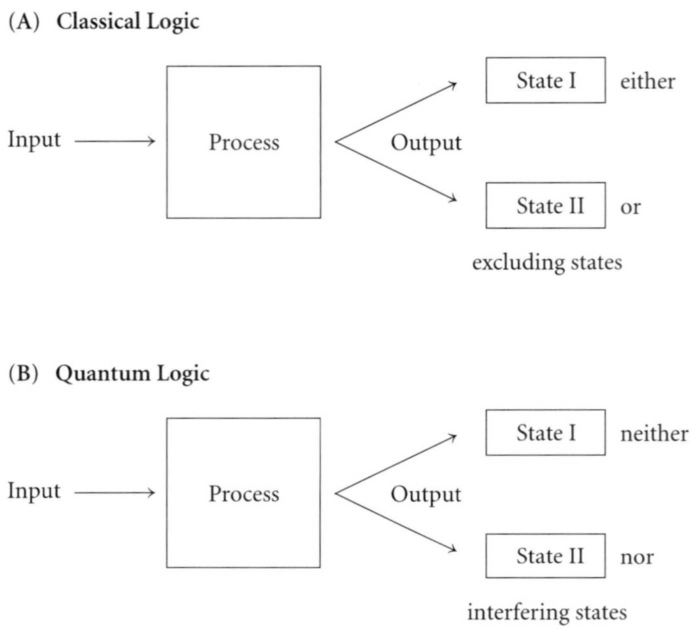

Figure 3. Schematic diagram of classical and quantum evolutions. (A) *excluding* states (classical determinism), (B) *interfering* states (quantum indeterminism).

by measurement. For simplicity, we restrict the distinguishable components of the output to only two states (Figure 3). In classical dynamics the output is *unique* (strict determinism), which means the result is *either* state I *or* state II: *excluding* states (Figure 3A). The very *essence* of a quantum process is, contrary to this, that the output is *not un*ique (no strict determinism), we have *neither* state I *nor* state II but a coherent superposition of both states: *interfering* states (Figure 3B).

In both cases the time development of the system is given by partial differential equations of first order in the time variable (Newton's or Maxwell's equations in the classical case, Schrödinger's equation in the quantum case) which describe the dynamics in a strictly *causal* way: the initial conditions determine *uniquely* the output. The *non-causal* element in the quantum case enters through the *von Neumann state collapse* (see previous paragraph) which occurs if one tries to realize the output state, either by a measurement, or by letting the output state undergo a successive process. Then the coherent superposition

$\alpha \cdot |\mathit{state\ I}\rangle + \beta \cdot |\mathit{state\ II}\rangle$
collapses into
either $|\mathit{state\ I}\rangle$ with probability $|\alpha|^2$,
or $|\mathit{state\ II}\rangle$ with probability $|\beta|^2$,
and $|\alpha|^2 + |\beta|^2 = 1$

For the *single event* – and it has to be emphasized that quantum mechanics is a theory for the single event, and not, as is sometimes claimed, an ensemble theory – the probabilities are completely irrelevant,[3] and the outcome is *completely unpredictable* (provided that not all but one of the probabilities are zero, which would imply the one left is equal to one). This constitutes the *non-computable* character of quantum events (Penrose 1994). It is evident that the deterministic logic underlying Cartesian dualism, which runs so heavily into conflict with the material world of classical physics, no longer applies if elementary quantum processes play a decisive role in brain dynamics.

3. Neocortical activity

Figure 4A illustrates the universally accepted six laminae of the neocortex (Szentagothai 1978) with two large pyramidal cells in lamina V, three in lamina III, and two in lamina II. The pyramidal apical dendrites finish in a tuft-like branching in lamina I (Figure 5A). There is agreement by Fleischhauer, Peters and their associates (Schmolke & Fleischhauer 1984; Peters & Kara 1987) that the apical bundles, diagrammatically shown in Figure 5B, are the basic anatomical units of the neocortex. They are observed in all areas of the cortex that have been investigated in all mammals, including humans. It has been proposed that these bundles are the cortical units for reception (Eccles 1990), which would give them a preeminent role. Since they are composed essentially of dendrites, the name *dendron* was adopted.

Figure 4B illustrates a typical spine synapse that makes an intimate contact with an apical dendrite of a pyramidal cell. The ultrastructure of such a synapse has been intensively studied by Akert and his associates (Pfenninger et al. 1969; Akert et al. 1975). The inner surface of a bouton confronting the synaptic cleft (d in Figure 4B, the active site in Figure 6A) forms the presynaptic vesicular grid (PVG) (Figure 6A–E). Figure 6B is a photomicrograph of a tangential section of a PVG, showing the dense projections in triangular array, and with the

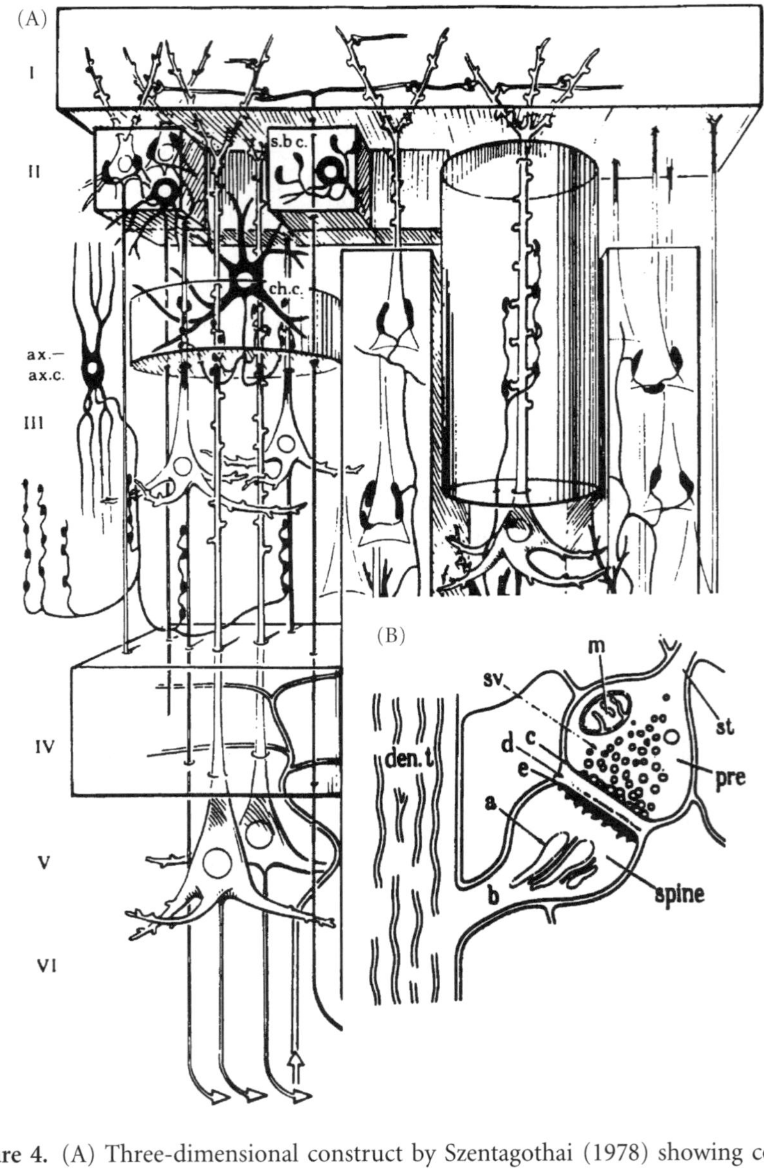

Figure 4. (A) Three-dimensional construct by Szentagothai (1978) showing cortical neurons of various types. There are two pyramidal cells in lamina V and three in lamina III, one being shown in detail in a column to the right, and two in lamina II. (B) detailed structure of a spine synapse on a dendrite (den.t); st, axon terminating in synaptic bouton or presynaptic terminal (pre); sv, synaptic vesicles; c, presynaptic vesicular grid (PVG in text); d, synaptic cleft; e, postsynaptic membrane; a, spine apparatus; b, spine stalk; m, mitochondrion (Gray 1982).

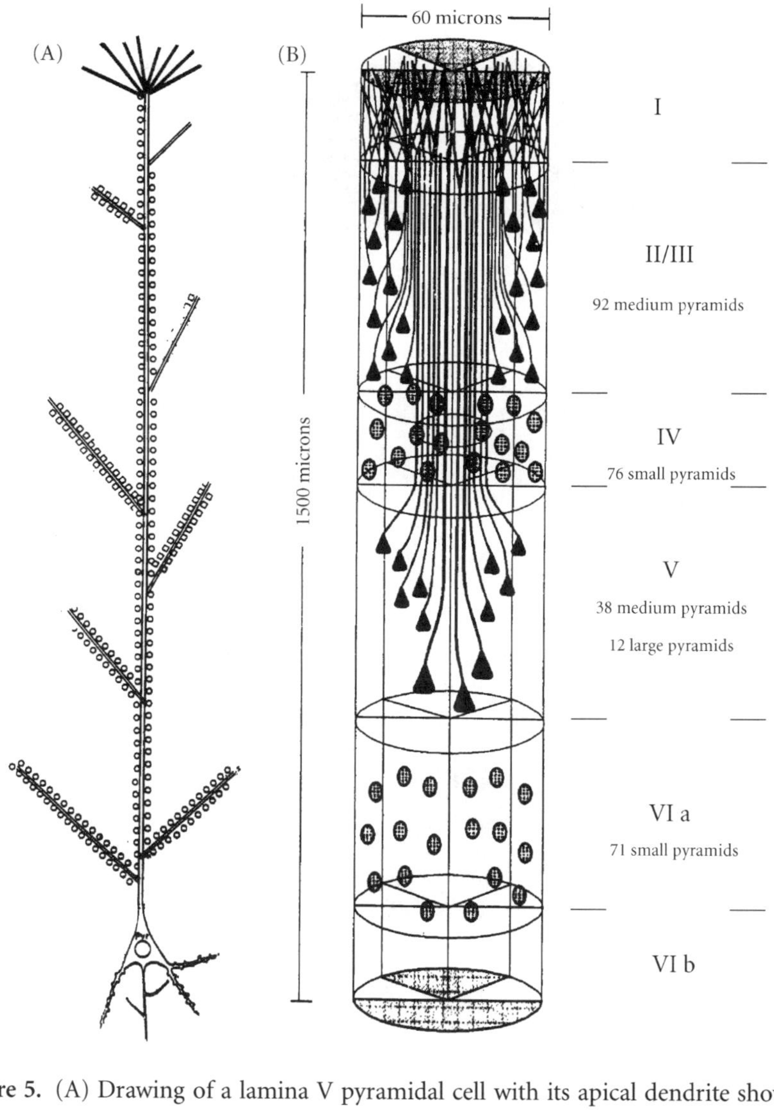

Figure 5. (A) Drawing of a lamina V pyramidal cell with its apical dendrite showing the side branches and the terminal tuft, all studded with spine synapses (not all shown). The soma with its basal dendrites has an axon with axon collateral before leaving the cortex. (B) Drawing of the six laminae of the cerebral cortex with the apical dendrites of pyramidal cells of laminae II, III and V, showing the manner in which they bunch in ascending to lamina I, where they end in tufts (Beck & Eccles 1992).

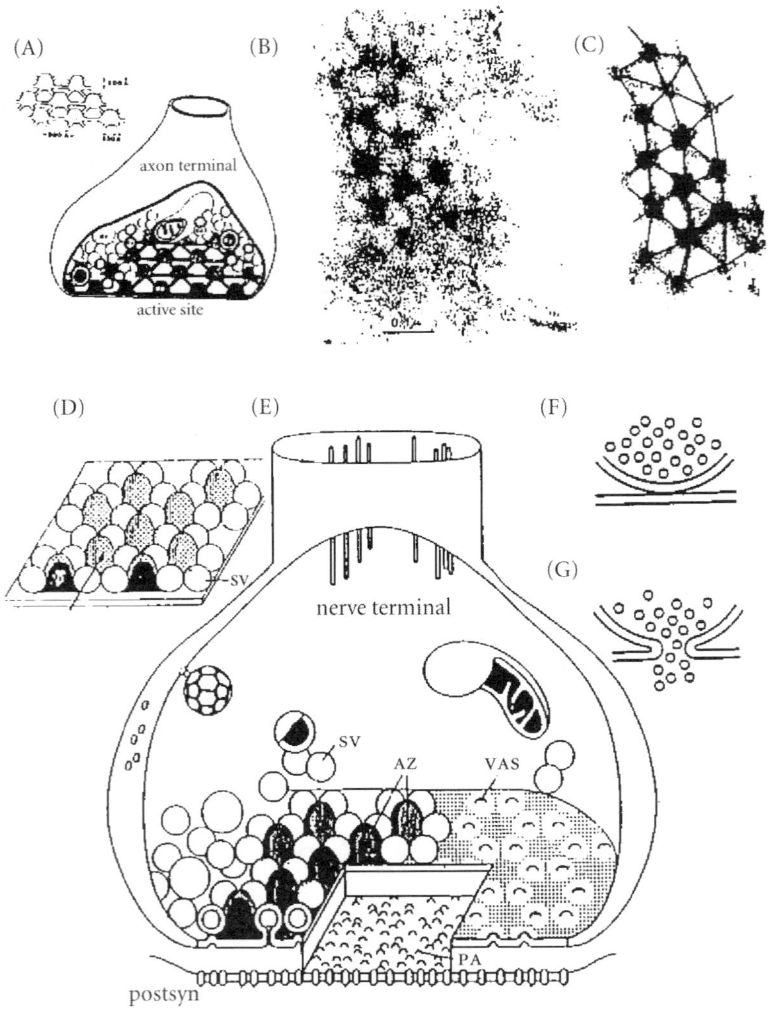

Figure 6. (A) Scheme of a nerve terminal, or bouton, showing the active site with cross linkages forming the PVG, which is drawn in Inset. (B, C) Tangential section through the presynaptic area. (D–E) Active zone (AZ) of mammalian central synapse showing geometrical design. SV, synaptic vesicle; VAS, vesicle attachement site; PA, presynaptic area. (F) Synaptic vesicle in apposition. (G) Exocytosis (Pfenninger et al. 1969; Akert et al. 1975; Gray 1982; Kelly et al. 1979; modified in Eccles 1994).

faint synaptic vesicles fitting snugly in hexagonal arrray. The spherical synaptic vesicles, 50–60 Å in diameter, with their content of transmitter molecules, can be seen in the idealized drawings of the PVG (Figure 6C and D). They

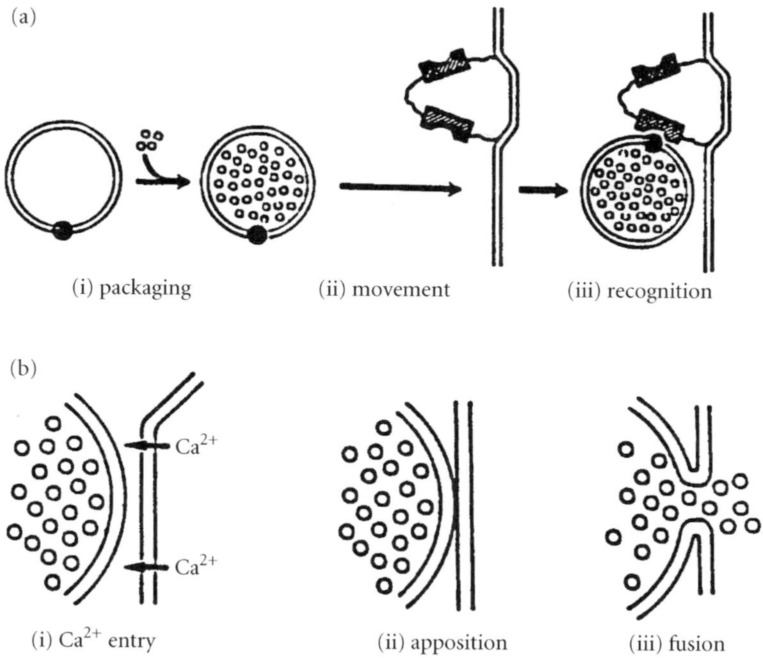

Figure 7. Different stages of synaptic vesicle propagation: (a) filling, movement towards the presynaptic membrane, docking. (b) stages of exocytosis. Note the essential role of Ca^{2+} after depolarization by a nerve impulse (Kelly et al. 1979).

arrange themselves in a hexagonal array on the active zone (Pfenninger et al. 1969; Akert et al. 1975).

A nerve impulse propagating into a bouton causes a process called *exocytosis*. At most, a nerve impulse evokes a single exocytosis from a PVG (Figure 6F and G, Figure 7). Exocytosis is the basic unitary activity of the cerebral cortex. Each all-or-nothing exocytosis of synaptic transmitter results in a brief excitatory postsynaptic depolarization (EPSP). Summation by electrotonic transmission of many hundreds of these milli-EPSPs is required for an EPSP large enough (10–20 mV) to generate the discharge of an impulse by a pyramidal cell (Figure 8). This impulse will travel along its axon to make effective excitation at its many synapses. This is the conventional macro-operation of a pyramidal cell of the neocortex, and it can be satisfactorily described by conventional neuroscience, even in the most complex design of neural network theory and neuronal group selection (Szentagothai 1978; Mountcastle 1978; Edelman 1989).

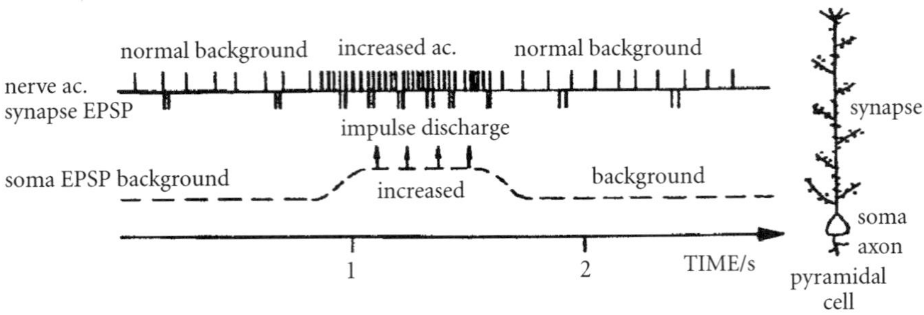

Figure 8. Impulse processing in a pyramidal cell (shown to the right). Upper trace: normal background and increased nerve activity. By filtering via synaptic exocytosis it produces reduced EPSPs, shown in the lower trace. The summed EPSP at the soma (excitation curve in the lower part) is not strong enough in normal background activity to produce an impulse discharge. This happens only with increased activity (Beck & Eccles 1992, sketch due to Helena Eccles).

Exocytosis has been intensively studied in the mammalian central nervous system, where it is meanwhile possible to refine the study by utilizing a single excitatory impulse to generate EPSPs in single neurons that are being studied by intracellular recordings. The initial studies were on the monosynaptic action on motoneurons by single impulses in the large Ia afferent fibres from muscle (Jack et al. 1981). More recently it was found that the signal-to-noise ratio was much better for the neurons projecting up the dorso-spino-cerebellar tract (DSCT) to the cerebellum.

This successful quantal resolution for DSCT neurons and motoneurons gives confidence in the much more difficult analysis of neurons of the cerebral cortex, which provide the key structures of neural events which relate to consciousness. The signal-to-noise ratio was so low in the studies of CA1 neurons of the hippocampus that so far only three quantal anlyses have been reliable in the complex deconvolution procedure by fluctuation analysis. In the most reliable case, a single axon of a CA3 hippocampal pyramidal cell set up an EPSP of quantal size 278 μV (mean value) in a single CA1 hippocampal pyramidal cell with approximately equal probabilities of release at each active site ($n = 5$) of 0.27 (Sayer et al. 1990). In the alternative procedure the single CA3 impulse projecting to a CA1 pyramidal cell was directly stimulated in the stratum radiatum. The EPSPs delivered by the deconvolution analysis of the two CA1 pyramidal cells were of quantal sizes 224 μV and 193 μV with probabilities ($n = 3$) of 0.24 and ($n = 6$) of 0.16, respectively (Sayer et al. 1989). For a systematic review, see Redman (1990). Key result of these observations is the fact that exocytosis can occur with probabilities much smaller than one per each impulse reaching the synapse.

4. Quantum versus classical brain dynamics

In the brain there exists an interplay between micro- and macrostructures. The latter consist of pyramidal cells, dendrites and their bundles (dendrons), electrochemical transitions, while microstructures involve synaptic membranes and microtubules. Nerve impulses, propagating along nerve cells are, independent of external stimuli or internal brain activity, always present and constitute a stochastic background in the brain. Recent investigations suggest that the neural net stays close to instability, and in this way can be switched by minute action between different states (Freeman 1996). In order to control such a system, a stable regulator has to be present which generates a coherent pattern in the active cortical unit. According to the cortical ultrastructure, as outlined in the previous section, synaptic action qualifies as this regulator. This has also been demonstrated in various biochemical studies of the influence of drugs and anesthesia on the ion channel properties of the synaptic membrane (Flohr 1995; Hameroff 1998). We argue in the following that because of the stochastic thermal background quantum action could only be effective in brain dynamics if it establishes itself as a *'quantum switch'* within the microstructures.

The all important regulatory function of spine synapses results from the fact that exocytosis, the release of transmitter molecules across the presynaptic membrane, occurs only with probabilities much smaller than one upon each incoming nerve impulse (Redman 1990). We therefore regard exocytosis as a candidate for quantum processes to enter the network, and thus regulating its performance (Beck & Eccles 1992).[4]

Micro- and macrostructures in the brain are clearly separated by time-, or correspondingly, energy scales. The macrostructure is typically characterized by the fact that the brain lives in *hot and wet* surroundings of $T \approx 300°K$. This raises immediately the question of quantum coherence vs. thermal fluctuations. As is well known, as soon as thermal energies surpass quantal energies classical thermal statistics prevails.

To answer this question, two characteristic energies can be defined:

i. The thermal energy per degree of freedom

$$E_{th} = \frac{1}{2} k_b T \quad \text{with } k_b\text{: Boltzmann's constant.}$$

ii. The quantal energy, defined as zero point energy of a quasiparticle of mass m_{eff} which is localized over a distance Δq. From Heisenberg's uncertainty relation $\Delta p \cdot \Delta q \geq 2\pi\hbar$ it follows (using the equal sign)

$$E_{qu} = \frac{(\Delta p)^2}{2M} \cong \left(\frac{2\pi\hbar}{\Delta q}\right)^2 \frac{1}{2m_{eff}} \quad \text{with } \hbar\text{: Planck's constant, devided by } 2\pi$$

These relations define *two energy regimes*

$E_{qu} << E_{th} = E_c$: the thermal regime
$E_{qu} >> E_{th}$: the quantal regime

with the breaking point E_c separating the two regimes which at physiological temperature of $T \approx 300°\text{K}$ amounts to an energy $E_c \sim 1.3 \times 10^{-2}$ eV.

Evaluation of the relation between the localization distance Δq and the effective mass m_{eff} shows that for moderate Δq of about 1–3 nm and effective masses below 10 m_e (m_e: electron mass) the quantal energy is well above the thermal regime $E < E_c$. This means that the dynamical mass of a quantum transition, if robust against thermal fluctuations, has to be of the order of a few electron masses. Biomolecules whose mass is in the range of kD, do not qualify *as a whole.*

We can also derive a *critical frequency*, $\hbar\omega_c = E_c$, and a *signal time*, $\tau_c = 2\pi/\omega_c$. With $E_c = 1.3 \times 10^{-2}$ eV one obtains

$$\omega_c \approx 2 \cdot 10^{13}\ \text{s}^{-1}; \qquad \tau_c \approx 0.3\ \text{ps}$$

These results show unambiguously that quantum processes at room temperature involve frequencies smaller than the picosecond scale. This, in turn, means they correspond to electronic transitions, like electron transfer or changes in molecular bonds (e.g., breaking of a hydrogen bridge).

Our analysis leads to the consequence that in brain dynamics two well separated regions with different time scales exist:

1. The *macroscopic*, or cellular, dynamics with time scales in the milli-, and down to the nanosecond range.
2. The *microscopic*, or quantal, dynamics with time scales in the pico- to femtosecond range.

The large difference in time scales makes it possible to deal with quantum processes in the individual microsites, and decoupled from the neural net. On the other hand, it explains why the usual biochemical and biophysical studies do not show the need for introducing quantum considerations. To uncover them one has to employ ultra-short time spectroscopy (Vos et al. 1993).

5. The quantum trigger model

Experimental analysis of transmitter release by spine synapses of hippocampal pyramidal cells has revealed a remarkably low exocytosis probability per excitatory impulse. This means that there exists an activation barrier against opening of an ion channel in the PVG. Activation can either occur purely stochastically by thermal fluctuations, or by stimulation of a trigger process. Here we propose a two-state quantum trigger which is realized by quasi-particle tunneling. This is motivated by the predominant role of exocytosis as the synaptic regulator of cortical activity which is certainly not completely at random. On the other hand, primary electron transfer processes play a decisive role in membrane transport phenomena (Vos et al. 1993).

Exocytosis as a whole certainly involves macromolecular dynamics (Figure 7). We propose, however, that it is initiated by a *quantum trigger mechanism*: An incoming nerve impulse excites some electronic configuration to a metastable level, separated energetically by a potential barrier $V(q)$ from the state which leads in a *unidirectional* process to exocytosis. Here, q denotes a *collective coordinate* representing the path along the coupled electronic and molecular motions between two states. The motion along this path is described by a *quasiparticle* of mass m_{eff} which is able to tunnel through the barrier quantum-mechanically. As has been shown in the previous section, m_{eff} can be in the range of tens of electron masses, or less, in order to survive thermal fluctuations. This implies that ion channel dynamics as a whole does not qualify for significant quantum processes in the brain.

The quasiparticle assumption allows the treatment of the complicated molecular transition as an effective one-body problem whose solution follows from the time dependent Schrödinger equation

$$i\hbar\frac{\partial}{\partial t}\Psi(q;t) = -\frac{\hbar^2}{2m_{eff}}\frac{\partial^2}{\partial q^2}\Psi(q;t) + V(q)\cdot\Psi(q;t).$$

Figure 9 shows schematically the initial state at $t = 0$ (after activation by the incoming impulse), and at the end of the activation period, $t = t_1$. Here it is assumed that the activated state of the presynaptic cell lasts for a finite time t_1 only before it recombines. t_1, however, is of the macroscopic time scale (micro- to nanosecond), as discussed in the previous paragraph. At $t = t_1$ the state has evolved into a part still sitting to the left of the barrier in region I, while the part in region III has tunneled through the barrier.

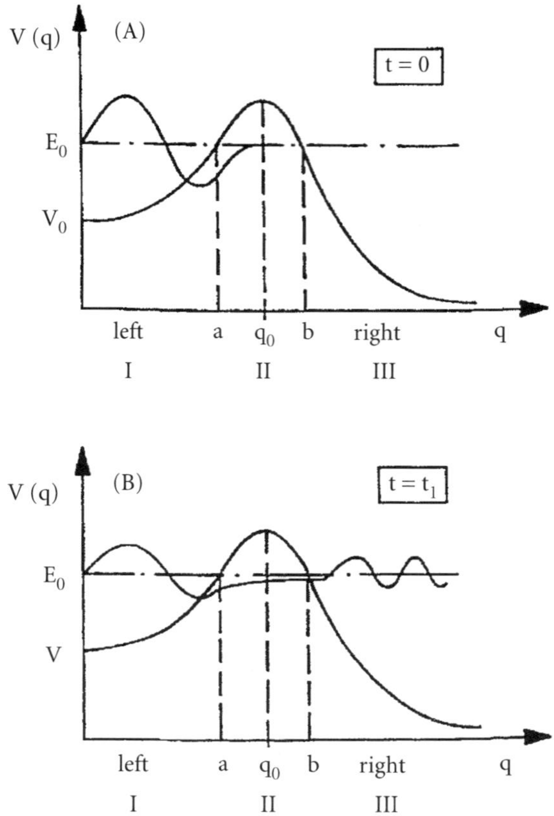

Figure 9. (A) The initial state (t = 0) of the quasiparticle in the potential V(q). The wave function is located to the left of the barrier. E_0 is the energy of the activated state which starts to tunnel through the barrier. (B) After time t_1 the wave function has components on both sides of the barrier. a, b: classical turning points of the motion inside and outside the barrier (Beck & Eccles 1992).

We can now separate the total wave function at time t_1 into two components, representing left and right parts:

$$\Psi(q; t_1) = \Psi_{left}(q; t_1) + \Psi_{right}(q; t_1),$$

and this constitutes the two *interfering amplitudes* for the *alternative results of the same process* as discussed in Section 2: *either* exocytosis *has happened* (Ψ_{right}), or exocytosis *has not happened* (Ψ_{left}) (inhibition). *State collapse* transforms this into

exocytosis: probability $p_{ex}(t_1) = \int |\Psi_{right}|^2 dq$
inhibition: probability $p_{in}(t_1) = \int |\Psi_{left}|^2 dq$

To once more estimating numbers using physically meaningful parameters we can evaluate the tunneling process using the Wentzel-Kramers-Brillouin (WKB) approximation (see, e.g. Messiah 1961). This results for the barrier transmission coefficient T in

$$T = \exp\left\{-2\int_a^b \frac{\sqrt{2m_{eff}[V(q) - E_0]}}{\hbar} dq\right\}$$

with E_0, the energy of the activated initial state. For barrier widths slightly above 1 nm and effective heights of 0.05 to 0.1 eV (which lies well above the energy of thermal fluctuations, cf. Section 4) one obtains transmission coefficients in the range 10^{-10} to 10^{-1}. Using an intermediate value of $T = 10^{-7}$, activation times t_1 in the macroscopic time scale of a few ns, and excitation energies (the energies of the activated quantum state before tunneling starts) between 0.5 and 1 eV, the probability for exocytosis $p_{ex}(t_1)$ covers the range between 0 and 0.7, in agreement with measured values (Jack et al. 1981). *Realistic numerical input leads to meaningful results for the quantum trigger which regulates exocytosis*!

As a consequence we can describe brain dynamics by two time scales:

microscopic scale (femtosecond):	*quantum synaptic dynamics*
macroscopic scale (nanosecond):	*(coherent) cell dynamics*
coupling:	*structure of the neuronal net*

As a *possible realization* we can consider *electron transfer* (ET) between biomolecules (Beck 1996b). In biological reaction centers such processes lead to *charge transfer* across *biological membranes* (Vos et al. 1993). The quasiparticle describes the electron coupled to nuclear motion according to the Franck-Condon principle. The theory has been worked out by Marcus (1956), and was later put into quantum mechanical version by Jortner (1976). The initializing step of ET is excitation of *donor* D, usually a dye molecule, with subsequent transport of an electron to *acceptor* A, producing the polar system D^+A^-. This is accompanied by rearrangement of the molecular coordinates leading to unidirectional charge separation and, over several further electronic transitions with increasing time constants, to the creation of an action potential across the membrane. The energetics is shown in Figure 10. Figure 10A shows the potential energy curves separately for electrons and nuclear conformations, Fig-

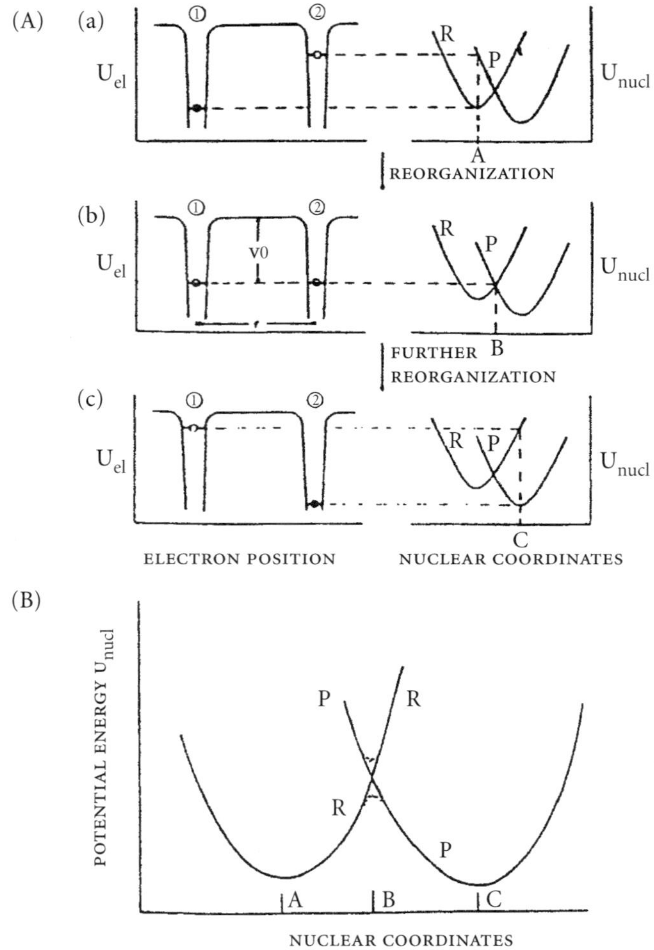

Figure 10. (A) Electron transfer coupled to nuclear motion. Left: electronic potential energy curves, right: corresponding nuclear potential curves. (a), (b), (c): electronic energies in the two wells for the nuclear positions A, B, C. The transition proceeds from (a) over the barrier (b) to the final state (c). (B) The same situation in the quasiparticle picture. The potential energy surfaces of donor (R) and acceptor (P) are shown. The positions correspond to (A), (B), (C) in (A). The dotted lines indicate splitting due to electron interactions between donor and acceptor (Marcus & Sutin 1985).

ure 10B gives the combined potential in the quasiparticle picture (Marcus & Sutin 1985). This latter form resembles closely the effective potential assumed in our quantum trigger model presented earlier in this section.

6. Neural coherence

Neural activity in processes of perception or intention is characterized by coherent action of specific areas in the brain (Singer 1990; Pardo et al. 1991; Eccles 1994). Activated areas are characterized by an increase in regional cerebral blood flow, as demonstrated in radio-xenon technology (Roland 1981), or more recently by positron emission tomography (PET, Posner et al. 1985; Corbetta et al. 1990). Activation generates most complex spatio-temporal patterns which characterize the specific perception (visual, audible, taste or touch) or intention (silent thinking, moves). These patterns are intimately related to memory and the learned inventory of pyramidal cells (Kandel & Schwartz 1982). In the neural bundles ('dendrons', cf. Figure 5) which comprise the active area, there are thousands of modifiable synapses which have to act cooperatively to generate the increased action potential needed to bring out the observed activity (Figure 8). Since the synapses can only modify (increase or decrease) their exocytosis probability upon incoming nerve impulses, there has to be a constant background activity which will be modulated *coherently* by a large number of synapses.

Long range coherence in biological systems at room temperature can either be established by self-organization in classical nonlinear dynamics, or by macroscopic quantum states (*Fröhlich coherence*, Fröhlich 1968). The necessary ingredients for macroscopic quantum states at room temperature are *dissipation* and *energy supply* (*pumping*). Pumping *stabilizes* against thermal fluctuations, and phase synchronization is achieved by *self organization*. The latter is mediated by nonlinear couplings to classical fields (electromagnetic, phonons, molecular) which implies that the quantal spectrum becomes *quasi-continuous*. Quantum state interference and subsequent state reduction, however, need a few well separated discrete states (like the two states in the synaptic trigger model), and consequently are not possible with macroscopic quantum states at room temperature.

From empirical evidence (Freeman 1996; Spitzer & Neumann 1996), and from successful modeling (Haken 1996), we would rather attribute long range cooperative action in the active zones of the brain to nonlinear dynamics of a driven open system. Such a system is far from thermal equlibrium and close to instability, and it can organize itself by external stimuli in a variety of synchronous activity patterns (Gray et al. 1989). Synaptic exocytosis in such a system serves as *regulator*, and the cooperation of the many synapses in the dendrons (active area) produce the spatio-temporal patterns above noise. Quantum action and subsequent state reduction in the individual synapse produce

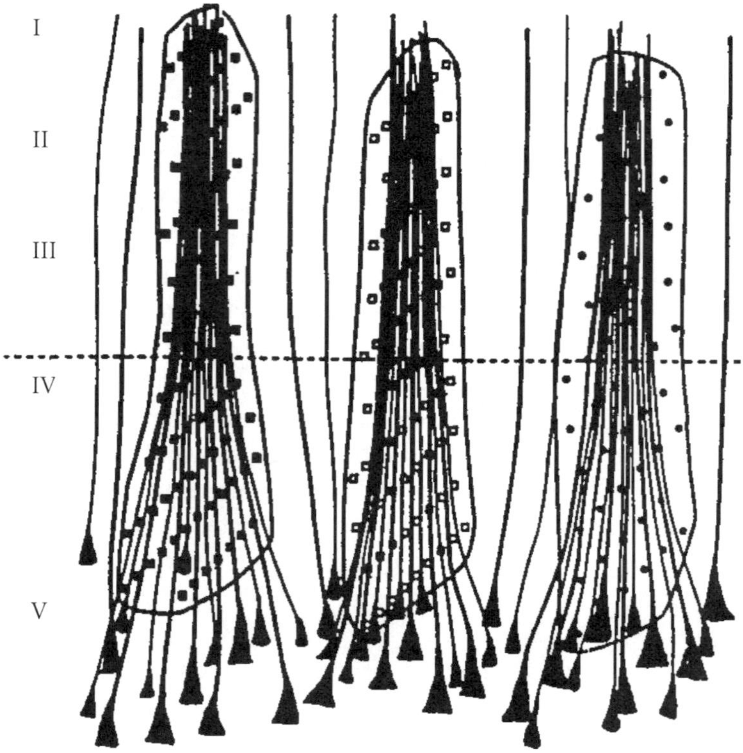

Figure 11. Coherent couplings of bundles of dendritic pyramidal cells (dendrons) to form spatio-temporal patterns (Eccles 1990).

the *non-algorithmic binding* in cortical units. Figure 11 presents a schematic sketch of three bundles of pyramidal cells (dendrons) surrounded by their spatial pattern which are produced temporarily by cooperation within the individual cells. Since these patterns are activated by *perception, intention* as well as in *ideation* (Ingvar 1989) they represent the basic *units of consciousness*. To give them a name which expresses their outstanding importance, we may call them *psychons* (Eccles 1990). The physiological mechanism of pattern formation and signal transduction in the brain are not yet fully understood. Recent rapid progress in understanding many facets of nonlinear dynamics in biological systems (Goldbeter 1996) gives, nevertheless, hope to proceed substantially in understanding large scale brain dynamics in the near future.[5] The important role of quantum events does, however, not depend on the exact nature of this large scale structure, but it relies on the concept of state superposition in microscopic molecular transitions.

Finally, a word concerning the *qualia* of consciousness may be added. Science can not, by its very nature, present any answer to the philosophical, ethical or religious questions related to the mind. It can, however, and it does, provide the *openness* which is essential to make discussions beyond the limitations of science possible. Thus the *'ignorabimus'* of Du Bois-Reymond (cf. Introduction) has been turned into *'non ignorabimus'* by quantum- and nonlinear physics!

7. Conclusions

Quantum *state collapse* is the decisive process which distinguishes quantum mechanics from classical physics. In a *single event* it is *non-predictable*. By this it qualifies for the indecisive and non-computable nature of brain functioning. It is emphasized that this introduces a new logical concept, different from classical determinism, underlying the struggle between dualism, identity theory, and the call for 'free will'. The interpretation of quantum mechanics as a succession of *single events* produces in a natural way the fundamental difference between past and future, in so far as the past is *known* (by events having manifestly occurred) while the future is *not known* (by the unpredictability of state reduction). One could, however, object against this interpretation by arguing that the Schrödinger equation is *causal*, and consequently describes the time evolution of the probability amplitudes unambiguously. Thus, the *probabilities* for future events are *completely determined*. This, however, relates to *ensemble averages*, for a large number of identical systems under identical initial conditions. Such ensembles can be realized in the material world of microscopic atomic systems but they are never realized in the world of complex objects such as the brain. Each new event finds itself borne in a new initial state. For these the non-predictability for single events prevails.

In view of these new and important concepts for elevating consciousness finally up to a scientific basis, we present evidence for a *realistic* implementation of quantum events into brain dynamics. It is based on our present knowledge of cortical structure and the synaptic regulation of neural impulses. Basic assumptions and results are:

– Quantum processes in the wet and hot surroundings of the brain are only possible at the microscopic level of (electron) transitions in the pico- to femtosecond time scale.

- Spine synapses are important regulators in brain activity, filtering the ever present firings of nerve impulses.
- Exocytosis, the release of transmitter substance across the presynaptic membrane, is an all-or-nothing event which occurs with probabilities much smaller than one.
- A model, based on electron transfer, relates exocytosis with a two-state quantum trigger, leading by quantum tunneling to the superposition of these two states, followed by state reduction (collapse into one definite final state).
- The coherent coupling of synapses via microtubular connections is still an open problem. Quantum coherence ('macroscopic quantum state') is not needed to couple microsites, which exhibit quantum transitions with their definite phase relations, to produce spatio-temporal patterns. The quantum trigger can however initialize transitions between different macroscopic modes (stochastic limit cycles, Grifoni & Hänggi 1996).
- The quantum trigger opens a doorway for a better understanding of the relations between brain dynamics and consciousness.

Notes

* This article was prepared by both authors in early 1997 and reviews their work on quantum aspects of brain dynamics. Shortly before finishing, Sir John Eccles passed away in May 1997. It is therefore his last publication and a memorial to Sir John and his lifelong struggle for understanding mind-brain interaction. Responsibility for the final text which has been slightly updated is, however, due to the first author.

1. In his book von Neumann describes the difference between the ensemble result and the single event rather drastically: "The everything leveling of the law of large numbers obscures completely the real nature of the single process".

2. One may argue that in this scenario the result depends on the place where the cut is made (e.g. in the above example at the screen or, alternatively, including the whole dynamics of the detectors). This is, however, not the case since cutting the dynamics at a later stage requires tracing over the non observed variables which makes the result independent of the cut.

3. Probabilities have relevance for expectations of frequencies in large series of *identical* events as, e.g., in coin tossing. For the outcome of a single event they say nothing. Brain processes are never series of *identical* events.

4. An alternative regulating process by tubulin molecules comprising the cylindrical walls of microtubules has been proposed by Hameroff and Penrose (1996). We would like to emphasize that the basic quantal event postulated by them, a two state conformational transition in the tubulin molecule, is rather similar to the synaptic quantum trigger model presented here. We do not follow, however, these authors in their postulate of a macroscopic coher-

ent quantum state along the microtubules, nor in their quantum gravitational arguing for '*orchestrated (state) reduction*' (OR).

5. A most promising approach to combine noisy structures with the enhancement of regular signals is presented by the observation of *stochastic resonance* (Gammaitoni et al. 1998). The combination of quantum tunneling states with a noisy surroundings has recently also been studied by Grifoni and Haenggi (1996).

References

Akert, K., Peper, K., & Sandri, C. (1975). Structural organization of motor end plate and central synapses. In P. G. Waser (Ed.), *Cholinergic Mechanisms.* New York: Raven.

Beck, F., & Eccles, J. C. (1992). Quantum aspects of brain activity and the role of consciousness. *Proceedings of the National Academy of Science USA, 89,* 11357–11361.

Beck, F. (1996a). Mind-brain interaction: Comments on an article by B. I. B. Lindahl & P. Århem. *Journal of theoretical Biology, 180,* 87–89.

Beck, F. (1996b). Can quantum processes control synaptic emission? *International Journal of Neural Sytems, 7,* 343–353.

Corbetta, M., Miezin, F. M., Dobmeyer, S., Shulman, G. L., & Petersen, S. E. (1990). Attentional modulation of neural processing of shape, color and velocity in humans. *Science, 248,* 1356–1359.

Descartes, R. (1644). *Principia philosophiae.* Amsterdam.

Donald, M. J. (1990). Quantum theory and the brain. *Proceedings of the Royal Society London, A427,* 43–93.

Du Bois-Reymond, E. (1916). *Über die Grenzen des Naturerkennens.* Leipzig: Veit & Co.

Eccles, J. C. (1990). A unitary hypothesis of mind-brain interaction in the cerebral cortex. *Proceedings of the Royal Society London, B240,* 433–451.

Eccles, J. C. (1994). *How the Self Controls Its Brain.* Berlin, Heidelberg, New York: Springer.

Edelman, G. M. (1989). *The Remembered Present: A Biological Theory of Consciousness.* New York: Basic Books.

Flohr, H. (1995). An information processing theory of anesthesia. *Neuropsychologia, 33,* 1169–1180.

Freeman, W. (1996). Random activity at the microscopic neural level in cortex ("noise") sustains and is regulated by low-dimensional dynamics of macroscopic cortical activity ("chaos"). *International. Journal of Neural Systems, 7,* 473–480.

Fröhlich, H. (1968). Long-range coherence and energy storage in biological systems. *International Journal of Quantum Chemistry, 2,* 641–649.

Gammaitoni, L., Haenggi, P., Jung, P., & Marchesoni, F. (1998). Stochastic resonance. *Review of Modern Physics, 70,* 223–287.

Goldbeter, A. (1996). *Biochemical Oscillations and Cellular Rhythmics.* Cambridge, UK: Cambridge University Press.

Gray, E. G. (1982). Rehabilitating the dendritic spine. *Trends in Neuroscience, 5,* 5–6.

Gray, C. M., König, P., Engel A. K., & Singer, W. (1989). Oscillatory response in cat visual cortex exhibit inter-columnar synchronization which reflects global stimulus properties. *Nature, 338,* 334–337.

Grifoni, M., & Hänggi, P. (1996). Coherent and incoherent quantum stochastic resonance. *Physical Review Letters, 76,* 1611–1614.

Gruber, H. E. (1974). *Darwin on Man* (p. 451). London: Wildwood House.

Haken, H. (1996). Noise in the brain: A physical network model. *International Journal of Neural Systems, 7,* 551–557.

Hameroff, S., & Penrose, R. (1995). Orchestrated reduction of quantum coherence in brain microtubules – a model for consciousness. In S. Hameroff, A. Kaszniak, & A. Scott (Eds.), *Toward a Science of Consciousness: Contributions from the 1994 Tucson Conference.* Cambridge, MA: MIT Press.

Hameroff, S. (1998). A unitary quantum hypothesis of anaesthetic action. In S. Hameroff, A. Kaszniak, & A. Scott (Eds.), *Toward a Science of Consciousness II: The 1996 Tucson Discussions and Debates.* Cambridge, MA: MIT Press.

Ingvar, D. H. (1990). On ideation and 'ideography'. In J. C. Eccles & O. Creutzfeldt (Eds.), *The Principles of Design and Operation of the Brain. Experimental Brain Research, Series 21.* Berlin, Heidelberg: Springer.

Jack, J. J. B., Redman, S. J., & Wong, K. (1981). The components of synaptic potentials evoked in cat spinal motoneurons by impulses in single group Ia afferents. *Journal of Physiology (London), 321,* 65–96.

Jortner, J. (1976). Temperature dependent activation energy for electron transfer between biological molecules. *Journal of Chemical Physics, 64,* 4860–4867.

Kandel, E. K., & Schwartz, J. H. (1982). Molecular biology of learning: modulation of transmitter release. *Science, 218,* 433–443.

Kelly, R. B., Deutsch, J. W., Carlsson, S. S., & Wagner, J. A. (1979). Biochemistry of neurotransmitter release. *Annual Review of Neuroscience, 2,* 399–446.

Lindahl, B. I. B., & Århem, P. (1994). Mind as a Force Field: Comments on a New Interactionistic Hypothesis. *Journal of theoretical Biology, 171,* 111–122.

Marcus, R. A. (1956). On the theory of oxydation-reduction reactions involving electron transfer. *Journal of Chemical Physics, 24,* 966–978.

Marcus, R. A., & Sutin, N. (1985). Electron transfer in chemistry and biology. *Biochimica Biophysica Acta, 811,* 265–322.

Margenau, H. (1984). *The Miracle of Existence.* Woodbridge, CT: Ox Bow.

Messiah, A. (1961). *Quantum Mechanics. Vol. I.* Amsterdam: North Holland.

Mountcastle, V. B. (1978). In F. C. Schmitt (Ed.), *The Mindful Brain* (pp. 7–50). Cambridge, MA: MIT Press.

Pardo, J. V., Fox, P. T., & Raichle, M. E. (1991). Localization of a human system for sustained attention by positron emission tomography. *Nature, 934,* 61–64.

Penrose, R. (1984). *The Emperor's New Mind: Concerning Computers, Minds and the Laws of Physics.* Oxford: Oxford University Press.

Penrose, R. (1994). *Shadows of the Mind: An Approach to the Missing Science of Consciousness.* Oxford: Oxford University Press.

Peters, A., & Kara, D. A. (1987). The neuronal composition of area 17 of the rat visual cortex. IV. The organization of pyramidal cells. *Journal of Comparative Neurology, 260,* 573–590.

Pfenninger, K., Sandri, C., Akert, K., & Eugster, C. H. (1969). Contribution to the problem of structural organization of the presynaptic area. *Brain Research. 12,* 10–18.

Popper, K. R. (1972). *Objective knowledge: An evolutionary approach.* Oxford: Clarendon Press.

Popper, K. R., Lindahl, B. I. B., & Århem, P. (1993). A discussion of the mind-brain problem. *Theoretical Medicine, 14,* 167–180.

Posner, M. I., Petersen, S. E., Fox, P. T., & Raichle, M. E. (1985). Localization of cognitive operations in the human brain. *Science, 240,* 1627–1631.

Redman, S. J. (1990). Quantal analysis of synaptic potentials in neurons of the central nervous system, *Physiological Review, 70,* 165–198.

Roland, P. E. (1981). Somatotopical tuning of postcentral gyrus during focal attention in man. A regional cerebral blood flow study. *Journal of Neurophysiology, 46,* 744–754.

Sayer, R. J., Redman, S. J., & Anderson, P. (1989). Amplitude fluctuations in small EPSPs recorded from CA1 pyramidal cells in the guinea pig hippocampal slice. *Journal of Neuroscience, 9,* 845–850.

Sayer, R. J., Friedlaender, M. J., & Redman, S. J. (1990). The time-course and amplitude of EPSPs evoked at synapses between pairs of CA3/Ca1 neurons in the hippocampal sclice. *Journal of Neuroscience, 10,* 626–636.

Schmolke, C., & Fleischhauer, K. (1984). Morphological characteristics of neocortical laminae when studied in tangential semi-thin sections through the visual cortex in the rabbit. *Anatomical Embryology, 169,* 125–132.

Singer, W. (1990). Search for coherence: a basic principle of cortical self-organization. *Concepts of Neuroscience, 1,* 1–26.

Spitzer, M., & Neumann, M. (1996). Noise in models of neurological and psychiatric disorders. *International Journal of Neural Systems, 7,* 355–361.

Squires, E. J. (1988). The unique world of the Everett version of quantum theory. *Foundations of Physics Letters, 1,* 13–20.

Stapp, H. P. (1991). Brain-mind connection. *Foundations of Physics, 21,* 1451–1477.

Szentagothai (1978). The neuron network of the cerebral cortex: A functional interpretation. *Proceedings of the Royal Society London B201,* 219–248.

von Neumann, J. (1955). *Mathematical Foundations of Quantum Mechanics, Chapter IV.* Princeton, NJ: Princeton University Press.

Vos, M. H., Rappaport, F., Lambry, J. C., Beton, J., & Martin, J.-L. (1993). Visualization of coherent nuclear motion in a membrane protein by femtosecond spectroscopy. *Nature, 363,* 320–325.

Wigner, E. P. (1964). Two kinds of reality. *Monist, 48,* 248–264.

Chapter 10

Quantum consciousness

A cortical neural circuit

Stuart R. Hameroff and Nancy J. Woolf
The University of Arizona / University of California

1. The problem of consciousness and the failure of conventional approaches

Perplexing features of consciousness

Despite ever-increasing knowledge of brain activities related to cognition, the mystery of conscious experience looms deeper than ever. How is it that neuronal activities give rise to thoughts, feelings, and free will (if it is indeed free)? How do neurotransmitters, membrane depolarizations and neural network dynamics produce the redness of a rose, or the feeling of happiness? Like the "Myth of Sisyphus", strenuous efforts to approach the problem through brain imaging and other techniques often leave us further and further from understanding the big picture. We are left scratching our heads, rather than comprehending what goes on inside them. The perplexing features of consciousness which induce this conundrum include:

- The nature of subjective experience or qualia – our "inner life" (Nagel 1974; Jackson 1994), Chalmers' "hard problem" (Chalmers 1996)
- Subjective binding of spatially distributed brain activities into unitary objects in vision, and a coherent sense of self or oneness
- Transition from pre-conscious processes to consciousness
- Non-computability or the notion that consciousness involves a factor which is neither random nor algorithmic, and that consciousness cannot be simulated by conventional computers (Penrose 1989, 1994, 1996)
- Free will

Of these, the problem of subjective experience is the most difficult. Philosophers call the raw components of our conscious experience "qualia". Why do we have qualia? Why do we have an inner life? Why isn't the world populated by unfeeling automaton "zombies" lacking qualia? Can qualia occur in computers? What specific aspect of brain function produces conscious experience?

Conventional approaches: Emergence

To approach these issues, modern science most commonly portrays consciousness as an emergent property of computer-like activities in the brain's neural networks. Since the work of Santiago Ramon-y-Cajal a century ago, the brain has been viewed as a large group of individual neuronal cells that communicate by chemical synapses involving various neurotransmitters. In the past half-century, the rise and utility of silicon computers have engendered comparisons between brain function and the computer, with neuronal firings and synaptic transmissions playing the role of information states, or bits in a computer. Neural network computer systems have emulated the parallelism and adjustably weighted synaptic connections found in the brain, achieving self-organization, perception, learning and other functions. These developments have reinforced the belief that the brain is a computer, and prompted speculation and predictions that consciousness will emerge in technological computational devices.

The basis for this approach is emergence. In nonlinear dynamics (e.g., chaos theory) emergence implies that a specific novel property occurs or emerges at some critical level of complexity in a hierarchical system dependent on activities at both lower and higher levels of organization. The brain is viewed as a hierarchical system, comprised of layers of organization with bottom-up, as well as top-down feedback. In this view, consciousness emerges as a novel property of computational complexity at an upper level of the hierarchy from nonlinear interactions among layers. Because novel properties can indeed emerge from complex interactions among simple components in a variety of systems (e.g., wetness from water, music or hurricanes from vibrations of air molecules), the inference is that consciousness emerges as a novel property of complex interactions among relatively simple neurons.

However there are significant problems with this conjecture. To begin, there is a complete lack of evidence for the emergence of consciousness (or cognitive functions) strictly from membrane and synaptic events. Among the many emergent phenomena in nature outside the brain (from a simple flame to the Great Red Spot on Jupiter), none appear to be conscious. Further, no testable predictions are suggested – no critical level or threshold of neuronal complexity is identified for the emergence of consciousness. By force-fitting the brain to a computer analogy, potentially important biological details that don't match the analogy are ignored. These details include widespread apparent randomness at all levels of neural processes and the potential roles played in cognition by glial cells, gap junctions between neurons and glial cells, dendritic processing, microtubules and microtubule-associated proteins (e.g., MAP-2), and metabotropic receptors

Apparent randomness of neural events

Randomness appears to be rampant throughout the brain. Approximately 15% of axonal firings that reach presynaptic terminals result in neurotransmitter release with a seemingly random distribution. Electrophysiological recordings typically include a great deal of background drift or noise commonly eliminated by taking the average of many recordings. Much activity is discarded. Arieli et al. (1996) recorded background ongoing activity simultaneously in various areas of mammalian brain and found that it correlated across brain regions. Perhaps noise represents unrecognized nonlocal signaling or information (Ferster 1996).

Glial cells

Approximately 80% of brain cells are glia, assumed in conventional approaches to be primarily insulation. However, glia have excitable membranes, ion channels, receptors, cytoskeletal microtubules, and are connected to neurons and other glia by gap junctions. It may be an oversight to assume that glia are not involved in consciousness.

Gap junctions between neurons and glial cells

In addition to chemical synapses, gap junctions connect neurons with other neurons, neurons with glial cells and glial cells with other glial cells. Gap junctions are window-like portholes, or pores of the protein connexin between adjacent nerve cells. Cytoplasm may flow through gap junctions that separate adjacent processes by only 4 nanometers. Nerve cells connected by gap junctions are electrically coupled and fire synchronously, behaving like one giant neuron.

Gap junctions are generally considered more primitive connections than chemical synapses, essential for embryological development but fading into the background in mature brains. Recent evidence, however, suggests that brain gap junctions may be important in the adult. In the adult, gap junctions are mainly found between glial cells or between neurons and glia (Nadarajah & Parnavelas 1999). Gap junction-connected neuronal networks may mediate coherent 40 Hz oscillations (Galarreta & Hestrin 1999; Gibson et al. 1999). These tight junctions can form between processes very rapidly, for example becoming suddenly more prevalent in brains undergoing acute withdrawal from amphetamines (Onn & Grace 2000). As will be discussed later, gap junctions could be important for macroscopic spread of quantum states among neurons and glia.

Dendritic processing

Dendrites are highly complex, branching structures where much of "the real work of the nervous system takes place" (Stuart et al. 1997). Dendrites are major neuronal compartments, accounting for over 90% of large pyramidal cells found in the hippocampus and neocortex (Larkman 1991; Stockley et al. 1993). Neuronal dendrites are frequency studded with dendritic spine protuberances that receive thousands of synaptic inputs. In the conventional approach, dendrites serve mainly to collect and funnel inputs via dendritic membrane depolarization to the neuronal soma and axon to determine axonal firing at the axon hillock.

It is not entirely clear, however, that the conventional approach to neuronal processing portrays the whole picture. A number of interesting and potentially important dendritic phenomena are ignored. For example, dendrites do a great deal towards shaping and integrating synaptic inputs (e.g., Crook et al. 1998). Back propagation of action potentials into the dendrites can occur and this may have important functional consequences (Buzsáki & Kandel 1998; Sourdet & Debanne 1999). Dendrites can also communicate directly with one another, generating dendritic spikes on occasion (see Schwindt & Crill 1998). Dendrodendritic synapses are common in olfactory structures (Sassoè-Pognetto & Ottersen 2000); however, synaptic connections between dendrites are only occasionally found in neocortex (Burchinskaya 1981). In the hippocampus, networks of dendrodendritic connections exist between GABAergic interneurons; 24% of these were identified as gap junctions (Fukuda & Kosaka 2000). Dendrodendritic gap junctions are common during development and are frequently found in brain regions that retain a high degree of plasticity in the adult, for example, the olfactory bulb (Isaacson & Strowbridge 1998). Immunohistochemistry for dendritic lamellar bodies provides indirect evidence for gap junctions in the hippocampus and cerebral cortex, as well as in the olfactory bulb (DeZeeuw et al. 1995), and direct electron microscopic evidence exists for gap junctions in the hippocampus of rats 8–12 week old (Fukuda & Kosaka 2000).

In addition to these novel modes of communication between dendrites, the static view of dendrites is misleading. Dendrites are continuously altering their shape, and the modifications of the outermost branches are in the highest state of flux.

The conventional view primarily focuses on one particular type of neural interaction: synaptic transmission from the axon to the somatodendritic compartment. However both Eccles (1992) and Pribram (1991) have advocated dendrodendritic processing as the primary site for consciousness.

Microtubules

Dendrites (and axons) have shapes determined by their interior cytoskeleton. Similar to the interiors of all living cells, neurons are functionally organized by webs of protein polymers-the cytoskeleton whose major components are microtubules. Structurally, microtubules are crystalline cylinders whose walls are hexagonal lattices of subunit proteins known as tubulin.

Microtubules are essential for a variety of biological functions including cell movement, cell division (mitosis) and establishment and maintenance of cell form and function. In neurons, microtubules and microtubule-associated proteins, such as MAP-2, self-assemble to extend axons and dendrites and form synaptic connections; microtubules then help maintain and regulate synaptic strengths responsible for learning and cognitive functions. (For a more complete description of the role of microtubules and other cytoskeletal structures in cognitive functions see Dayhoff et al. 1994; Hameroff & Penrose 1996a; Hameroff 1994.)

While microtubules have traditionally been considered as purely structural components, recent evidence has demonstrated mechanical signaling and communication functions (Glanz 1997; Maniotis et al. 1997a, 1997b; Vernon & Wooley 1995). Microtubules interact with membrane structures and activities by linking proteins (e.g., fodrin, ankyrin) and second messenger signal operating through G-proteins.

As will be described in Section 2, theoretical models suggest that microtubules process information in both classical and quantum computational modes, and furthermore that such processes are essential to consciousness.

Metabotropic receptors

Many computational models of the brain are based simply on excitation versus inhibition, yet in reality we have a many neurotransmitters and even more receptors. The conventional view considers the functional effects of excitatory neurotransmitters on receptors to be entirely related to effects on dendritic membrane depolarization, and indeed ionotropic receptors (e.g., nicotinic acetylcholine, AMPA and kainate glutamate receptors) do act primarily to depolarize the postsynaptic dendritic membrane. However, receptors termed metabotropic (e.g., muscarinic acetylcholine receptors, metabotropic glutamate receptors, serotonergic receptors, D1-D5 dopamine receptors, etc.) have significant effects on the cytoskeleton through second messengers mediated through the activation of G-proteins.

Following metabotropic receptor activation, changes in the phosphorylation state of various cytoskeletal proteins occur. This affects microtubule-MAP-2 networks, where one might expect these chemical cascades to lead to structural changes in dendrites and to participate in conscious events (Woolf 1997). This will be discussed further in Sections 3 and 4.

Cognition in single cells

The computational emergence approach considers neuronal firings and synaptic events as simple states, essentially information bits of either "1" or "0". However single cell organisms such as paramecium swim, avoid obstacles, find food, mates, have sex, and learn, escaping from capillary tubes more quickly with subsequent trials. As single cells they have no synapses – they utilize membrane and especially cytoskeletal microtubules as their nervous system. If a protozoan can swim, find food and reproduce, wouldn't a single cell neuron be more capable than acting as a simple switch? Processing within the neuron, for example in the cytoskeleton, might well be relevant to cognition and consciousness.

If not emergence, then what?

Although the computational emergence approach has been useful, we believe that in terms of understanding consciousness, strict adherence to such a view does not lead to a satisfactory solution. The conventional synaptic approach provides no obvious endpoint nor does it provide a framework for considering various neurobiological processes in the assimilation of consciousness. The problem of consciousness demands that we consider all details of brain function and structure, and probe the limits of available science. But if consciousness is not an emergent property of complex computer-like neuronal activities, what is it?

A theory of consciousness must (1) account for the perplexing features, (2) be consistent with neuroscientific evidence including often ignored messy details, and (3) generate testable predictions. In this paper we describe a model of brain function extending to the quantum realm, which can meet these criteria.

2. Quantum computation and consciousness – how quantum approaches can explain perplexing features of consciousness

Quantum computation

Consciousness and other brain functions have historically been compared to contemporary information processing technologies. In ancient Greece memory was likened to a "seal ring in wax", and more recently brain and mind have been likened to telegraph circuits, holograms and silicon computers. The next vanguard of information processing is quantum computation, which relies on effects of quantum mechanics to process information.

Quantum computers were described theoretically in the 1980's (e.g., Benioff 1982; Feynman 1986; Deutsch 1985) to utilize the phenomenon of quantum superposition, in which a particle can exist in two or more states or locations simultaneously (see below). Whereas current computers represent information as bits of either "1" or "0", quantum computers are proposed to utilize quantum bits – called qubits – of superposition of both "1" AND "0". Potential advantages of quantum computing stem from multiple qubits in superposition interacting nonlocally (by quantum coherence or entanglement, see below); this implements near-infinite quantum parallelism. These interactions perform computations, and at some point, the qubits collapse, or reduce to a single set of classical bits – the solution. Other types of quantum computation involve quantum interference patterns and quantum annealing of spin-network systems. Quantum computers offer significant technological and commercial advantages, for example quantum computers are potentially able to rapidly factor large numbers into their primes, revolutionizing cryptography. Prototype quantum computers now exist and research activity is intense.

When quantum computers become a dominant technology, comparisons between brain, mind and quantum computation will be inevitable. We contend that such comparisons will be valid, and are valid now, for if certain types of quantum effects can govern brain dynamics, perplexing features of consciousness can be explained.

Quantum mechanics and collapse of the wave function

Quantum mechanics describes how particles and energy behave at small scales. Surprisingly, quantum particles act as both particles and waves, which can exist in quantum coherent superposition of two or more states or locations simultaneously. Despite the illogical nature of the situation, experiments have repeatedly shown that an atom, ion and even a molecule as large as a 60-carbon fullerene may exist in quantum superposition, separated from itself by 80 nanometers or more.

Another odd feature of quantum particle/waves is quantum entanglement. If two quantum particles are coupled but then go separate ways, they remain somehow connected over space and time. Measurement of one will instantaneously affect the state of the other, no matter how far away. Also, particles can condense into one collective quantum object in which all components are governed by the same wave function (e.g., Bose-Einstein condensate). Another odd feature is that, at the quantum level, time ceases to exist, or at least it becomes irrelevant. Reality at the quantum level is indeterminate, and extremely difficult to explain.

Consider quantum superposition. How can an object, no matter how microscopic, be in two or more states or locations simultaneously, and why don't we see this bizarre condition in our macroscopic "classical" world? At the quantum level, behavior of wave-like objects in superposition can be satisfactorily described in terms of a deterministic process (e.g., state vector) evolving according to the Schrödinger equation. On the other hand large-scale (classical) systems seem to obey different deterministic laws and to be in definite states and locations. The nature of the boundary between the quantum and classical worlds is a mystery, related to what is known as the "measurement problem", or "collapse of the wave function" in quantum mechanics.

A century of experimentation has indicated that when quantum systems are measured or observed, they cease to exist in superposition and choose particular classical states or spacetime locations (a particular eigenstate – one state of many possible states). This process is known in various contexts as collapse of the wave function, quantum jump, Heisenberg event or quantum state reduction. Von Neumann (1966), Schrödinger (1935) and other early quantum theorists supposed that quantum collapse effectively occurred when a quantum system interacted with its environment, including being measured or consciously observed, with the choice of particular post-collapse eigenstates being random. Similarly, modern theories suggest that quantum systems decohere due to environmental interactions. The boundaries between the quantum and classical realms are not completely clear, however.

Some viewpoints do not take collapse as a real phenomenon. For example in the approach of David Bohm (see Bohm & Hiley 1993), wave functions are guided by "pilot waves" from a deeper reality, and apparent collapses involve local minima in a complex field. In the "multiple worlds" view of Everett (1957), Deutsch (1985) and others, the multiple possibilities in a superposition avoid collapse by each leading to its own separate realities – a multitude of universes. Others take collapse, or reduction to be an actual physical process.

Penrose "objective reduction: OR" and fundamental spacetime geometry

A number of physicists have argued in support of schemes in which the rules of quantum mechanics are modified by the inclusion of some additional procedure, which induces collapse, or reduction due to some objective threshold or feature. These schemes (objective reduction: OR) include those suggesting reduction occurs due to a critical number of superpositioned particles ($\sim 10^{17}$: Ghirardi et al. 1986; cf. Pearle 1989), and some which consider gravitational effects (e.g., Károlyházy et al. 1986; Diósi 1989; Ghirardi et al. 1990; Penrose 1989; Pearle & Squires 1994). Penrose (1994) describes OR in which quantum superposition persists until it reaches a critical threshold related to quantum gravity and then abruptly self-collapses. According to Penrose, the threshold for objective reduction or self-collapse in a quantum system isolated from environment is given by the indeterminacy principle, $E=\hbar/T$. E is the energy of the mass in superposition, $\hbar$ is Planck's constant over 2π, and T is the time until reduction occurs. One may determine E by the amount of mass in superposition and the distance of separation of the mass from itself. Thus according to Penrose OR, large systems in isolated superposition will collapse quickly, whereas small systems may persist. For example, see Table 1 (from Penrose 1994; Hameroff 1998a):

Table 1.

E	T
nucleon (proton or neutron)	10^7 years
beryllium ion	10^6 years
tubulin protein (10^5 nucleons)	2 years
water droplet ($r = 10^{-5}$ cm)	2 hours
water droplet ($r = 10^{-4}$ cm)	50 msec
2×10^{10} tubulin proteins	25 msec (interval of 40 Hz)
water droplet ($r = 10^{-3}$ cm)	0.001 msec
Schrodinger's cat (1 kg)	10^{-37} sec

The Penrose OR approach has a number of advantages in regards to quantum mechanics and the measurement problem. It generates testable predictions, relies on a fundamental (indeterminacy) principle, and connects the process to quantum gravity, a description of fundamental reality. This latter connection may have relevance to consciousness, and it unifies quantum mechanics and general relativity (i.e., Einstein's gravity). One additional important feature proposed by Penrose is that the choices of classical eigenstates after each OR event is not random, nor entirely deterministic, but reflective of an additional non-computable influence occurring at the precise moment of collapse.

This "non-computability", Penrose argues, is a subtle but important feature of consciousness, a clue to its underlying mechanism that suggests that OR may be occurring in the brain. The non-computable influences are embedded in the fundamental level of spacetime geometry in which the OR process is occurring, according to Penrose, and reflect "the Platonic realm" of mathematical truth, as well as perhaps aesthetic values and qualia. Where exactly is this realm? What is the fundamental level of reality? If the brain is organized hierarchically, perhaps consciousness extends downward to the deepest level of spacetime geometry.

Several approaches to the deepest structure of spacetime have emerged. One approach is string theory (superstrings) that states that quarks, electrons and other particles of matter, rather than being point-like, are actually tiny line-like objects called strings. The strings are each approximately the size of the Planck length (10^{-33} cm), the smallest possible distance in spacetime. It is known that at the Planck scale (10^{-33} cm, 10^{-43} sec) spacetime is no longer smooth, but quantized. The tiny strings in spacetime can supposedly vibrate in many ways to create the different types of elementary particles, and they potentially can explain the force-carrying particle/waves (e.g., bosons) that act on matter. Although mathematically elegant, string theory is currently untestable, and requires 11 dimensions. It is unclear whether these extra dimensions actually exist or are mathematical abstractions with no basis in reality.

Another approach to the fundamental nature of reality, which requires only 4 dimensional spacetime, is quantum geometry. To provide a description of the quantum mechanical geometry of space at the Planck scale, Penrose (1971) introduced "quantum spin networks" in which spectra of discrete Planck scale volumes and configurations are obtained. Smolin (1997) has generalized these spin networks to describe quantum gravity.

There are reasons to suspect gravity, and in particular quantum gravity in the fundamental makeup of spacetime geometry and as the basement level in consciousness. According to Einstein's theory, gravity is unique in physics because it is the only physical quality which influences causal relationships between spacetime events, and because gravitational force has no local reality, as it can be eliminated by a change in spacetime coordinates; instead, gravitational tidal effects provide a curvature for the very spacetime in which all other particles and forces are contained. It follows from this that gravity is a fundamental component of physical reality.

It may seem surprising that quantum gravity effects could plausibly have relevance at the physical scales relevant to brain processes. Quantum gravity is normally viewed as having only absurdly tiny influences at ordinary dimensions. However, according to the principles of Penrose OR this is not the case, and scales determined by basic quantum gravity principles are indeed relevant for conscious brain processes.

Quantum superposition and spacetime separation

According to modern accepted physical pictures, reality is rooted in 3-dimensional space and a 1-dimensional time, combined together into a 4-dimensional spacetime. This spacetime is slightly curved, in accordance with Einstein's general theory of relativity, in a way that encodes the gravitational fields of all distributions of mass density. Each mass density affects a space-time curvature, albeit tiny.

To envision such curvature, 4-dimensional spacetime may be conveniently portrayed schematically as a 2-dimensional "spacetime sheet", with one dimension of space and one dimension of time (Figure 1). Mass is equivalent to curvature in spacetime, so a mass in one location may be represented by curvature out of the picture (towards the viewer), and the mass in a second location as a curvature into the picture (away from the viewer). The two different spacetime curvatures in the top of Figure 1 represent mass (this could be a tubulin protein) in two different locations or conformations respectively. Mass in quantum superposition is simultaneous spacetime curvature in opposite directions, a separation, or bubble in spacetime, (e.g., bottom, Figure 1). Therefore according to Penrose, quantum superposition is a very slight separation, bubble or blister in fundamental spacetime geometry.

This is similar to the picture in the "multiple-worlds" view, in which superposition is a separation of underlying reality. In "multiple worlds" small separations invariably grow and lead to separate universes, or realities. However in the Penrose approach, the spacetime separations are unstable. At a critical degree of separation, the system must select either one state or the other – hence OR occurs. There remains only one universe. Critical spacetime separation is given as described earlier by $E=\hbar/T$, applicable to both mass separation and its equivalent spacetime separation. Thus Penrose OR is a self-organizing process at the dynamic boundary between the quantum world and classical world, accessing and influenced by the fundamental level of spacetime geometry. If

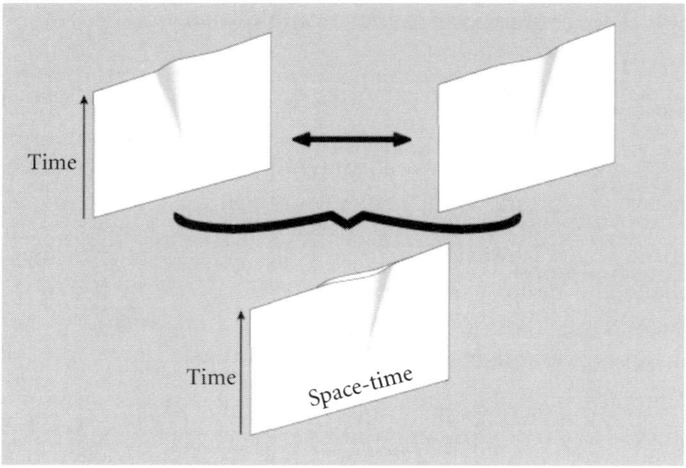

Figure 1. Quantum coherent superposition represented as a separation of space-time. The top two diagrams illustrate bifurcating space-time, in which two alternative mass distributions exist in quantum superposition. In the lowest of the three diagrams, a bifurcating space-time is depicted as the union ("glued together version") of the two alternative space-time histories that are depicted at the top of the figure (adapted from Penrose 1994: 338).

proto-conscious qualia and non-computable Platonic values are indeed embedded at the fundamental level, then a brain process accessing and amplifying the Planck scale could explain perplexing features of consciousness.

What is needed, therefore, is a form of quantum computation which collapses to solutions by Penrose OR and which governs brain functions. Technological quantum computers are envisioned to reduce to classical output solutions by environmental interactions or measurement. As the qubits in technological quantum devices are likely to involve electron spins, or perhaps atoms, the mass/spacetime separation energy E will be very small, and T will be exceedingly long. Technological quantum computers will not undergo OR and will not be conscious according to our approach. For a system to undergo OR in a time T relevant to brain processes (e.g., tens to hundreds of msec), then nanograms of mass in the brain must be superposed/separated.

A number of sites and various types of quantum interactions have been proposed; however, we strongly favor microtubules as an important component. As mentioned earlier, microtubules are assemblies of proteins known as tubulin. The next question is can proteins act as qubits?

Protein conformational dynamics

Dynamical regulation of neuronal activity depends on conformational states of many proteins including ion channels opening or closing, receptors changing shape upon binding of neurotransmitter, enzymes, second messengers and cytoskeletal signaling and movement. How are protein conformational states regulated? Proteins in a living state are dynamic, with transitions occurring at many time scales, however conformational transitions in which proteins move globally and upon which protein function generally depends occur in the nanosecond (10^{-9} sec) to 10 picosecond (10^{-11} sec) time scale (Karplus & McCammon 1983). Amazingly, proteins are only marginally stable. A protein of 100 amino acids is stable against penetration by only ~ 40 kiloJoules per mole (kJ mol^{-1}) whereas thousands of kJ mol^{-1} are available in a protein from side-group interactions including van der Waals forces. Consequently protein conformation is a "delicate balance among powerful countervailing forces" (Voet & Voet 1995).

The types of forces operating among amino acid side groups within a protein include charged interactions such as ionic forces and hydrogen bonds, as well as interactions between dipoles – separated charges in electrically neutral groups. Dipole–dipole interactions are known as van der Waals forces and include three types: (1) permanent dipole – permanent dipole, (2) permanent dipole – induced dipole and (3) induced dipole – induced dipole.

Type 3 (induced dipole – induced dipole) interactions are the weakest but most purely non-polar. They are known as London dispersion forces, and although quite delicate (40 times weaker than hydrogen bonds), are numerous and influential. The London force attraction between any two atoms is usually less than a few kilojoules; however, thousands occur in each protein, particularly within non-polar regions of proteins called "hydrophobic pockets". As other forces cancel out, London forces in hydrophobic pockets can govern protein conformational states.

London forces ensue from the fact that atoms and molecules that are electrically neutral and spherically symmetrical nevertheless have instantaneous electric dipoles due to asymmetry in their electron distribution. The electric field from each fluctuating dipole couples to others in electron clouds of adjacent non-polar amino acid side groups. Due to inherent uncertainty in electron localization, London forces are quantum mechanical effects whose disruption in a set of brain proteins causes anesthesia (e.g., Franks & Lieb 1992; Hameroff 1998a).

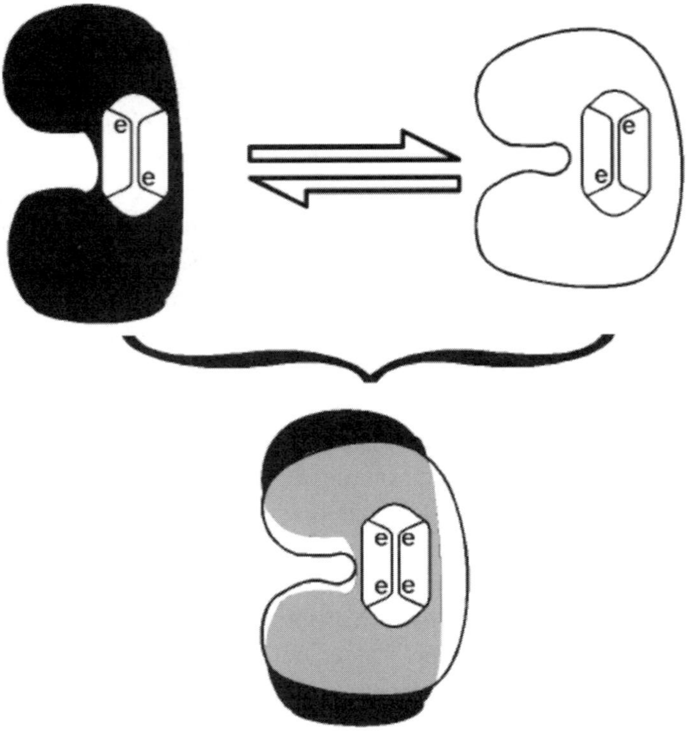

Figure 2. Schematized diagram of tubulin protein switching between two conformational states governed by van der Waals interactions in a hydrophobic pocket. (Tubulin may actually have several smaller collectively governing hydrophobic pockets.) Top: Tubulin switching between 2 conformational states coupled to localization of paired electrons (London force) within a hydrophobic pocket. Bottom: Quantum superposition (simultaneous existence in two distinct states) of the electron pair and protein conformation.

Quantum dipole oscillations within hydrophobic pockets were first proposed by Frohlich (1968) to regulate protein conformation and engage in macroscopic coherence, and Conrad (1994) suggested quantum superposition of electrons leads to superposition of various possible protein conformations before one is selected. Roitberg et al. (1995) showed functional protein vibrations that depend on quantum effects centered in two hydrophobic phenylalanine residues, and Tejada et al. (1996) have evidence to suggest quantum coherent states exist in the protein ferritin. Tubulin is a barbell-shaped dimer with a large hydrophobic region (Figure 2), and seemingly a suitable qubit device.

Microtubules and the "Orch OR" model

Theoretical models propose that microtubule subunit tubulins undergo coherent excitations, for example, in the gigaHz range by a mechanism suggested by Frohlich called pumped phonons (see Frohlich 1968, 1970, 1975; cf. Penrose & Onsager 1956). Frohlich excitations of tubulin subunits within microtubules have been suggested to support computation and information processing (e.g., Hameroff & Watt 1982; Rasmussen et al. 1990) by clocking computational transitions occurring among neighboring tubulins acting as cells as in molecular scale cellular automata. Dipole coupling among neighboring tubulins in the microtubule lattice acts as a transition for simulated microtubule automata exhibiting information processing, transmission and learning.

Classical microtubule automata switching in the nanosecond scale offers a potentially huge increase in the brain's computational capacity. Conventional approaches focus on synaptic switching (roughly 10^{11} brain neurons, 10^3 synapses/neuron, switching in the millisecond range of 10^3 operations per second) and predict about 10^{17} bit states per second for a human brain (e.g., Moravec 1988). However as biological cells typically each contain approximately 10^7 tubulins (Yu & Bass 1994), nanosecond switching in microtubule automata predicts roughly 10^{16} operations per second, per neuron. This capacity could account for the adaptive behaviors of single cell organisms like paramecium, for example, which elegantly swim, avoid obstacles, and find food and mates without benefit of a nervous system or synapses. As the human brain contains about 10^{11} neurons, nanosecond microtubule automata offer about 10^{27} brain operations per second.

However, even a vast increase in computational complexity won't by itself address the difficult issues related to consciousness. Quantum coherent or entangled states and quantum computation with objective reduction could possibly do so. Full rationale and details of the Orch OR model are given in Penrose and Hameroff (1995) and Hameroff and Penrose (1996a, 1996b). Key points are listed in Section 4.

The Orch OR model can account for the enigmatic features of consciousness, based on a set of assumptions, and generates testable predictions. In the next section we consider how such cytoplasmic quantum processes can fit in the context of known neuroscience.

3. How quantum computation can operate in the brain

Orch OR in microtubules is a proposal for a quantum process occurring within the neuronal and glial cytoplasm. How could such an apparently delicate process be leveraged to exert appropriate volitional influence, and to experience qualia, in the context of what is presently known about brain structure and function? We postulate three types of neural circuitry in cerebral cortex provide the basis for consciousness and its attention-related modulation through the Orch OR mechanism (Figure 3).

- *Thalamocortical inputs.* Thalamic inputs relay specific sensory information to each vertical column of cortex. These inputs provide nonconscious, neurophysiological information about a sensory field, mapped in a point-to-point fashion by glutamatergic synapses. Only in conjunction with memory and attention (see below), can spatial and temporal patterns among these inputs induce conscious, coherent quantum wave functions.
- *Basalocortical inputs.* Acetylcholine transmitted from the basal forebrain increases attention and enables conscious awareness (Sarter 1999; Perry et al. 1999; Woolf 1999a). Our view is that cholinergic activation enables selected spatial and temporal patterns provided by thalamocortical inputs to induce conscious, coherent quantum wave functions. This is hypothesized to occur via metabotropic receptor activation leading to isolation of microtubules and Orch OR.
- *Gap junction syncytia.* Coherent cortical oscillations in the EEG gamma frequency (30–70 Hz activity, also known as "40 Hz") appear to be critical for consciousness (Singer 1999). Recent evidence suggests inhibitory GABAergic cortical interneurons are important mediators of 40 Hz activity (Galarreta & Hestrin 1999; Gibson et al. 1999). These interneurons are interconnected by gap junctions, and form "dual" connections with each pyramidal dendrite: an inhibitory GABA chemical synapse and an electrotonic gap junction connection (Tamas et al. 2000).

The cerebral cortex and its connections: Thalamocortical inputs

Currently, the leading anatomical candidates for activities most directly related to consciousness involve forebrain regions detailed exquisitely by Ramon-y-Cajal (1909). These include circuits in the cerebral cortex and thalamus, which are essentially the continuation of sensory pathways from the retina, tectum, sensory nuclei of the brainstem and spinal cord. Fibers from the thalamus

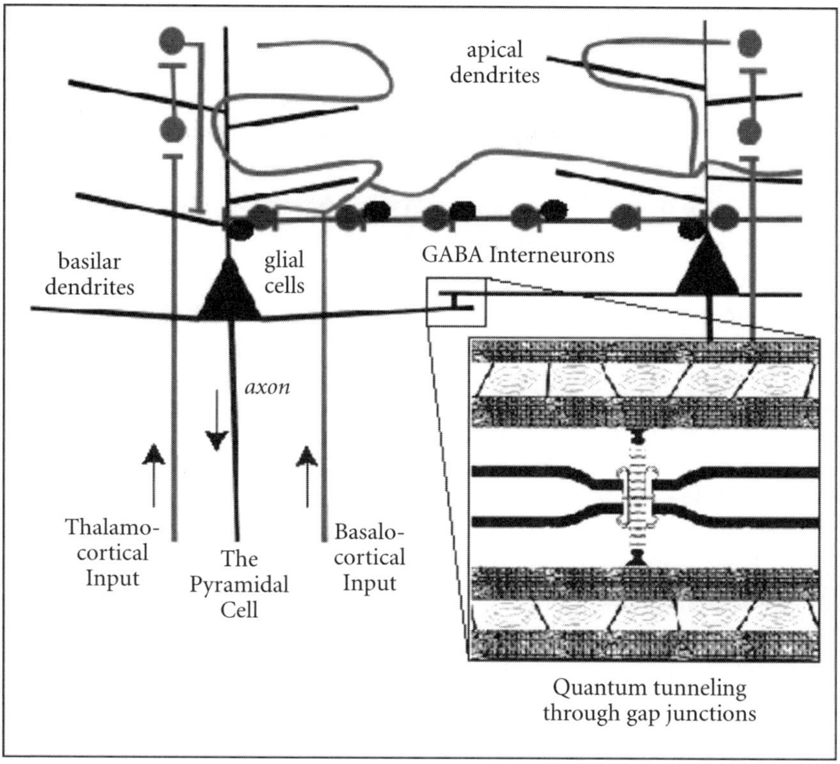

Figure 3. The basic cortical circuit in consciousness. The pyramidal cell is the central character, receiving thalamocortical and basalocortical inputs that initiate, respectively, classical and quantum computation. Quantum coherence and superposition are entangled among adjacent cortical circuits by transient gap junction connections involving (1) basilar dendrites of pyramidal cells and (2) gap junctions linking GABA interneurons and glial cells. An enlargement of the box showing gap junctions between the basilar dendrites illustrates quantum tunneling through the gap junction. This gap junction links two dendritic cytoplasms transmitting coherent quantum activity in the microtubules (horizontal bars) tuned by MAP-2 proteins (vertical lines).

make an initial synapse in layer 4 of cerebral cortex; then small local circuit cells relay that information up and then down, finally reaching the large pyramidal cells. These sensory circuits primarily use the fast acting neurotransmitter glutamate. Pioneering work by Herbert Jasper in the 1960's (e.g., Jasper & Komaya 1968) showed that sensory input passes through thalamus where it is broadcast to cortex. Besides thalamocortical projections that carry specific sensory modalities, other pathways are non-specific, but necessary for arousal and consciousness.

In recent decades, downward projections from cortex to thalamus have been described, and a consensus view holds that reverberatory feedback (recurrent loops) between thalamus and pyramidal neurons in cortex provides the neural correlate of consciousness (e.g., the "global workspace" of Baars 1988). Electrophysiological recordings further reveal coherent firing of these thalamocortical loops with frequencies varying from slow EEG frequencies (2–12 Hz) to rapid gamma oscillations in the 40 Hz range and upward. Coherent gamma frequency thalamocortical oscillations (collectively known as coherent 40 Hz) are suggested to mediate temporal binding of conscious experience (e.g., Singer et al. 1990; Crick & Koch 1990; Joliot et al. 1994; Gray 1998). Thalamocortical 40 Hz activity stands as a prevalent view of the neural level substrate for consciousness.

Nonetheless, thalamocortical loops may not be as important to consciousness as are activating systems, such as the ascending reticular formation and the cholinergic basal forebrain. Neurons in the basal half of the brain look different from most other neurons – almost like cells in tissue culture with sweeping dendrites weaving lawlessly through fiber bundles and adjacent brain areas. Connections are more global than serial (see Woolf 1996), and occasional gap junctions link dendrites together. These reticular neurons primarily use acetylcholine and monoamine neurotransmitters, which most often act on metabotropic receptors. This contrasts with neurons in the thalamus and cerebral cortex, typically using glutamate and usually exerting a rapid action at AMPA or Kainate receptors.

These activating systems originate from ventral or basal regions of the brain as opposed to the dorsally located cerebral cortex and thalamus. The dorsal-ventral division is most pronounced in early neural development, yet throughout life, the ventral or basal half of the central nervous system is organized entirely different from the dorsal part. Basal neurons capture the vision of Camillo Golgi insofar as these neurons are reticular, having dendrites entangled into a net. Reticular neurons are found in virtually all basal regions, including the basal forebrain, basal ganglia, hypothalamus, raphe nuclei, substantia nigra, locus ceruleus and reticular formation. These types of neurons found in the basal forebrain are discussed next.

The cerebral cortex and its connections: Basal forebrain inputs

The basal forebrain receives cholinergic and monoaminergic projections from the brainstem (i.e., the pedunculopontine nucleus, laterodorsal tegmental nucleus, locus ceruleus, raphe nuclei, and substantia nigra) Before reaching the basal forebrain, these brainstem neurons mingle with ascending sensory pathways, sampling incoming data of different sensory modalities. These neurons then send axons, which travel through the thalamus (partly through the so-called "intralaminar thalamus") and then reach the basal forebrain. In the case of noradrenergic and serotonergic neurons, axons also reach the cerebral cortex directly. The cholinergic basal forebrain system – or the nucleus basalis of Meynert as it is called in the human – is the site of degeneration in Alzheimer's disease. There are at least 500,000 cholinergic cells in each of the right and left halves of the basal forebrain, which send axons throughout the cerebral cortex, hippocampus, and amygdala. Large pyramidal cells and their vast dendrite arbors are the most common targets (Figure 4). Together, neurons from basal forebrain and brainstem modulate the activity of the cerebral cortex by increasing or decreasing the release of acetylcholine, norepinephrine, serotonin, and dopamine near to the large pyramidal cells.

One feature of the cholinergic basal forebrain system is that neurons lying close to one another often reach widely separated, discrete regions of cerebral cortex. For example, individual modules of cerebral cortex measuring 1–2 mm^2 (in the rat) are each innervated by separate and distinctly "devoted" cholinergic neurons. This contrasts with noradrenergic inputs to cerebral cortex, which are very widespread, having individual axons branching out to several cortical regions. Hence the cholinergic basal forebrain seems to be the most specific modulator of neocortex and limbic cortex, with monoaminergic projections complementing cholinergic modulation and playing general arousal roles of their own. Acetylcholine also appears to dominate as a cortical neuromodulator based on sheer amounts present. There is roughly 10 times as much acetylcholine in cerebral cortex as serotonin, and 20 times as much cortical acetylcholine as there is norepinephrine (Molinengo & Ghi 1997). Anesthetic drug studies, dream studies, and studies of aberrant states of consciousness in Alzheimer's dementia uniformly indicate that the cholinergic basal forebrain plays a prominent role in consciousness (Perry et al. 1999).

Thus, there are two types of cortical inputs with contrasting features. In our view, serial thalamocortical neurons relay sensory information to cortex as faithfully as possible, without directly accessing consciousness. On the other hand, global afferents, such as highly interconnected basal forebrain neurons

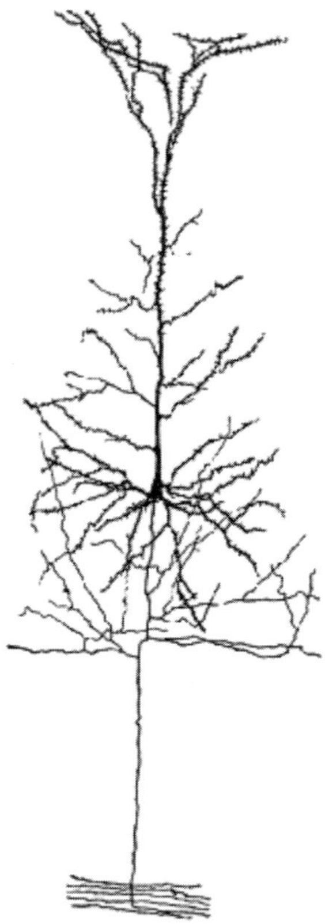

Figure 4. Cortical pyramidal cell demonstrated by the method of Golgi. Figure adapted from Ramon-y-Cajal (1909).

modulate or highlight particular sets of dendrites receiving sensory data that can be successfully fit to a coherent quantum wave function, thereby focusing attention and selecting specific items for consciousness through the Orch OR mechanism.

Quantum computing by the Orch OR mechanism is expected to largely occur deep within pyramidal cell dendritic cytoplasm, but nonetheless be linked to synaptic membrane events by second messenger chemical cascades activated via metabotropic receptors (Figure 5). Acetylcholine, operating through the metabotropic muscarinic receptor, activates the second messenger, phosphoinositide-specific phospholipase C (PI-PLC), which in turn activates protein kinase C and calcium/calmodulin-dependent kinase II (Siegel et al. 1998). These kinases then add phosphoryl groups to specific sites on the microtubule-associated protein-2 (MAP-2) molecule (Sanchez et al. 2000;

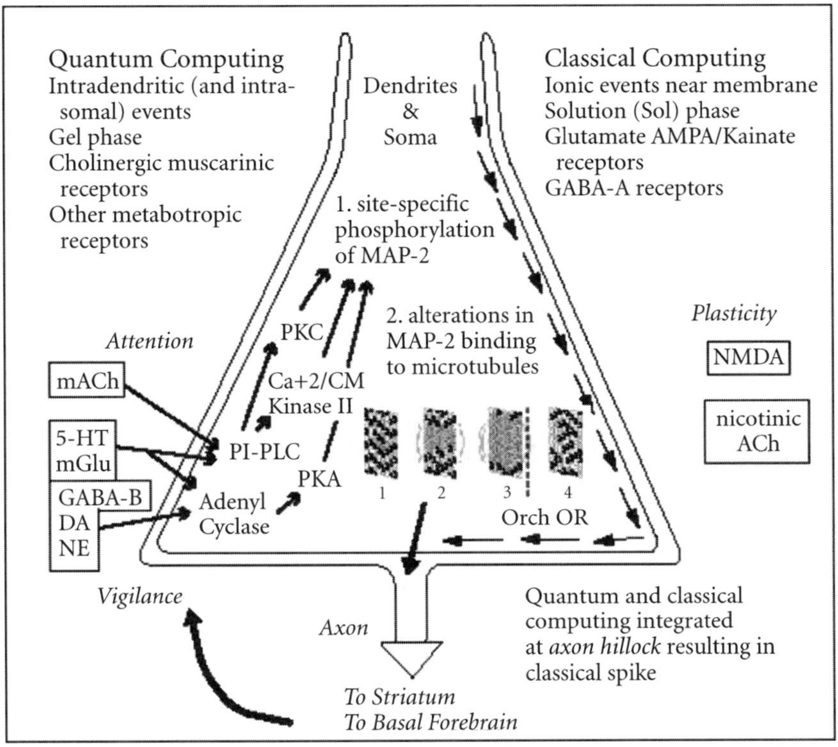

Figure 5. Quantum computing is accomplished by metabotropic receptor types, classical computing by ionotropic receptors. Abbreviations: calcium/calmodulin-dependent kinase II (C/CMK); dopamine receptors (DA); GABA receptors (GABA-B; GABA-A); ionotropic glutamate receptors (AMPA/Kainate); metabotropic glutamate receptors (mGlu); microtubule-associated protein-2 (MAP-2); muscarinic acetylcholine receptor (mACh); norepinephrine receptors (NE); orchestrated objective reduction (Orch OR); phosphoinositide-specific phospholipase C (PI-PLC); protein kinase A (PKA); protein kinase C (PKC); serotonin receptors (5-HT).

Johnson & Jope 1992). Serotonin, norepinephrine, glutamate and histamine, seemingly less than acetylcholine, activate PI-PLC via subpopulations of their own metabotropic receptors, leading to additional phosphorylation of MAP-2.

MAP-2 phosphorylation at these particular sites acts to decouple MAP-2 from microtubules and from actin, the net effect being to isolate microtubules from membrane and environmental influences (Figure 5). PI-PLC activation also directly isolates actin molecules from the neuronal membrane (Tall et al. 2000), further isolating microtubules embedded in actin gel. We suggest these mechanisms of isolation initiate quantum coherent wave functions in microtubules and then the Orch OR mechanism results in phenomenal consciousness. Each Orch OR event also regulates classical neural functions (Figure 5).

Horizontal gap junction syncytia – "Hyperneurons"

Glutamatergic circuits may be ideally suited to convey point-to-point sensory information, and cholinergic circuits may be ideal for initiating quantum coherence in single neurons. But how could quantum coherent states develop simultaneously in multiple cortical neurons? Difficulties have been posited (Tegmark 2000; Seife 2000). Many horizontally connected neurons (approximately10^5–10^7) transiently sharing a common cytoplasmic interior helps to overcome apparent difficulties and to permit widespread Orch OR (Woolf and Hameroff 2001; Hagan et al. 2001).

Laterally connected cortical pyramidal cells are cited as the most likely sites for consciousness (e.g., see Eccles 1992; Pribram 1991); however, the horizontal spread of activation has been considered limited: traditional axodendritic synapses simply do not extend very far laterally, at least not in any systematic way. There are, nonetheless, dendrodendritic and dendritic-glial connections via gap junctions, which can provide extensive and far-reaching lateral interactions. Gap junction-connected neurons fire synchronously, and are essentially one continuous membrane and one continuous cytoplasm. E. R. John (1986) termed gap junction networks "hyperneurons", and suggested they may reach across significant portions of the brain.

Gap junctions between dendrites in the brain have a specific intracellular organelle within the cytoplasm of each dendrite. Immediately adjacent to the gap junction pore are structures tethered to small cytoskeletal proteins anchored to microtubules. The structures consist of layers of membrane covering a mitochondrion, and they are called dendritic lamellar bodies (de Zeeuw 1995). Their function is unknown, but they have been suggested to facilitate quantum tunneling between dendrites. Such unity is important for a coherent quantum wave function to govern many separate neurons.

Making and breaking of gap junctions are regulated by microtubules. These gap junction connections may form quickly and be removed quickly. Thus, gap junction networks (hyperneurons) may be transiently present, lasting only hundreds of milliseconds. We suggest transient hyperneurons house unitary quantum states supporting Orch OR in their collective microtubules, and that particular configurations correspond with particular sequences of conscious events.

Gap junctions are relatively rare among neurons in comparison to chemical synapses. We would expect such interconnectedness to be sparse, maximizing the variety, number and specificity of possible configurations. Pyramidal cell dendrites receive thousands of chemical synapses; however, for a gap junction network, merely three connections per pyramidal cell dendrite would be sufficient to link extensive portions of the brain. Inclusion of glial cells would also greatly favor spread throughout significant brain volume. The transient nature and the topological geometry of hyperneurons make them appealing candidates for neural correlates of consciousness.

GABA interneurons could support Orch OR in three ways: (1) transiently quieting cortical membrane activity, thereby minimizing environmental decoherence; (2) synchronizing brain-wide, coherent 40-Hz activity (Desmedt & Tomberg 1994); and (3) enabling spread of cytoplasmic quantum states through gap junctions.

4. Applications of the model

Qualia – the nature of conscious experience

As described in Hameroff (1998b), an alternative to emergence is that consciousness is somehow fundamental, with qualia being primitive, fundamental aspects of reality, irreducible to anything else (e.g., like spin, or charge). Philosophical renditions along these lines have included panpsychism (e.g., Spinoza 1677), panexperientialism (e.g., Whitehead 1920) and most recently pan-protopsychism (Wheeler 1990; Chalmers 1996). In panpsychism all matter has some consciousness, whereas in panexperientialism and pan-protopsychism there exists a fundamental proto-conscious entity convertible to consciousness by some action of the brain. Perhaps most compatible with modern physics is Whitehead's "panexperientialism" that portrays consciousness as a sequence of discrete events (occasions of experience) occurring in a wider, proto-conscious field. Abner Shimony (1993) pointed out that Whitehead's discrete events were consistent with quantum state reductions. If we take Whitehead's "proto-conscious" field to be the fundamental level of reality itself, then quantum processes at this level could be related to consciousness. As it happens, quantum computers are illuminating such an approach.

Memory

Quantum entanglement is an excellent candidate process for associative memory insofar as it enables nearly limitless possibilities for which neural representations might be paired together. Moreover, a mechanism exists for permanently storing the way in which microtubules are "tuned" by MAP-2, which would be expected to influence the nature of Orch OR and the waveform of quantum events spread over a set of dendritic microtubules in neocortex or hippocampus. The mechanism for storing the way in which microtubular events are "tuned" is through degradation and subsequent structural rearrangement of MAP-2.

MAP-2 is a dendrite-specific microtubule-associated protein whose scaffolding activity is correlated with memory (Woolf 1998; Woolf et al. 1999b). According to the conventional view, synapses, not dendrites are the primary site of plasticity accompanying learning and memory. Learning and memory, nonetheless, appear to involve reorganization of proteins inside the dendrite, which may result in increased dendrite branching (see Woolf 1996, 1998; Woolf et al. 1999). Learning-related dendritic remodeling, some of which can be interpreted as postsynaptic responses accompanying synaptic plasticity and some of which are clearly not related to the synapse. In most cases, reorganization of microtubule and MAP-2 protein probably occurs some distance from the synapse, as we know that the synapse-laden spines are largely devoid of these proteins.

In our view, the structural modification of MAP-2 during memory encoding would be expected to alter conscious content as MAP-2 "tuning" of Orch OR would be modified whenever bridges between MAP-2 and microtubules are altered. Thus, MAP-2 and the signal transduction molecules that regulate its phosphorylation state may well be critical to memory that relies on conscious awareness, and awareness that relies on memory. These proposals are fully testable.

Visual consciousness

As elaborated in Woolf and Hameroff (2001), the Orch OR model can be successfully applied to the problem of visual consciousness. In our view, a sequence of increasingly greater intensity (and shorter duration) conscious events processes various individual aspects of visual information, integrating and enveloping (by gap junction connections) successively more entangled information and quantum superposition in more and more microtubule subunits and neurons resulting ultimately in a cumulative, high intensity conscious moment which integrates all aspects in a visual gestalt. Figure 6 illustrates this concept. Cholinergic inputs to cortical representations of visual input select a sequence of particular aspects: object shape may be identified in visual areas V2, V3, VP and LO, color in V8 and V4v, motion in V5, V3A and V7, and a cumulative visual gestalt in all of these and primary visual cortex (V1). The individual Orch OR conscious events processing the various aspects of the epoch (including the final gestalt) become more intense, with shorter durations, but have an average duration of around 25 msec; hence these oscil-

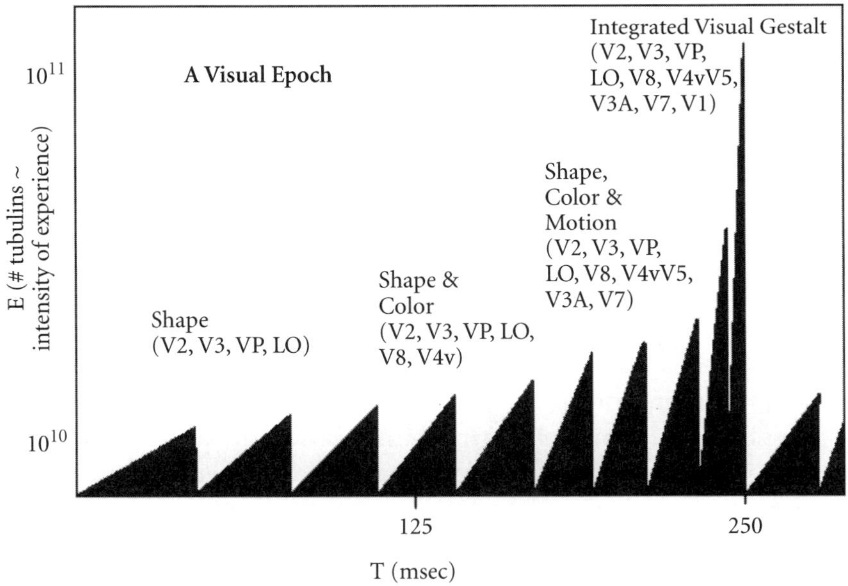

Figure 6. A crescendo sequence of ~25 msec quantum computations (i.e., oscillating near to 40 Hz) comprise a visual epoch lasting 250–700 msec. The time until Orch OR (threshold for conscious event) is given by the indeterminacy principle $E = \hbar/T$, where E is related to magnitude of the superposition, $\hbar$ is Planck's constant over 2π, and T is the time until self-collapse. Thus isolated superpositions that are large (high intensity, vivid experience) will reduce quickly, and isolated small superpositions (low intensity, bland experience) will require a long time. This implies a spectrum of conscious events of varying intensity and content. Phenomenal consciousness is suggested to correspond to putative and known functional roles of visual areas of cortex (see Logothetis 1999). Abbreviations: large-scale object visual cortex (LO); visual areas of cortex: primary, secondary, etc. (V1–V8); ventral posterior visual cortex (VP).

late at 40 Hz. Each crescendo sequence resulting in a visual gestalt is an epoch of several hundred milliseconds.

We postulate that thalamocortical inputs to visual cortex paint an entire non-conscious visual scene from which subsets of cortical neurons/aspects of the visual scene are selected for conscious attention by cholinergic inputs from basal forebrain. Inherent in this proposal is the prediction that perception will be better for visual stimuli that can be readily described by a wave function compared to those that cannot. This may indeed be the case, at least for some types of visual stimuli. Neurons in V1 respond better to sine gratings approximating a series of stripes rather than to highly contrasting black and white stripes (DeValois & DeValois 1988), and other kinds of periodic-pattern-selective cells have been identified in V1 (Von der Heydt et al. 1992).

Volition, choice, and free will

The steps in the Orch OR model ultimately culminate in an event that potentially accounts for volition, choice and free will. These steps are as follows:

1. Conformational states of individual tubulin proteins in brain microtubules are governed by internal quantum events (e.g., London forces in hydrophobic pockets) and able to cooperatively interact with other microtubule lattice tubulins in two types of phases: classical and quantum computation. Classical phase computation (in which microtubules act as automata) is done in communication with the external environment, receiving input from, and regulating output to, chemical synapses and other neural membrane activities. In this phase the cytoplasm surrounding and containing microtubules is in a liquid (solution or sol) state.
2. In the quantum phase, actin gelation, ordered water and a condensed charge phase surrounds, isolates and insulates microtubules from environmental decoherence. The gelation/quantum isolation phase may be triggered by synaptic influence through decoupling of microtubule-associated proteins, such as MAP-2, in response to increased phosphorylation (Woolf 1997). In the isolation phase microtubules are embedded in a gelatinous (gel) state, decoupled from external environment. Thus sol-gel transitions in cytoplasm, a basic activity in all of our cells, switches the microtubule environment between alternating phases of classical and quantum computation at frequencies consistent with neurophysiological membrane events (e.g., 40 Hz).

3. In the quantum (gel) phase, quantum coherent superposition emerges among London force electron pairs in hydrophobic pockets of microtubule subunit tubulins in microtubule lattices (similar to the proposal by Frohlich 1968, 1970, 1975). Superposition of London force electrons within each tubulin's hydrophobic pocket induces superposition of tubulin's atomic structure. Each tubulin becomes "separated from itself" at the level of each of its atom's nuclei, and alternate superpositions correspond (and lead) to alternate post-OR conformations. Each tubulin thus acts as a qubit, engaging in quantum computation with other quantum coherent state tubulins. In this phase, quantum computation among tubulins evolves linearly according to the Schrodinger equation (with microtubules acting as quantum automata).
4. Microtubule quantum superposition/computational states link to those in other neurons and glia by tunneling through gap junctions or quantum coherent photons traversing membranes (Jibu & Yasue 1995; Jibu et al. 1994, 1996). This enables spread of macroscopic quantum states in networks of transient gap junction-connected cells (both neurons and glia) throughout large brain volumes.
5. The proposed quantum superposition phase in neural microtubules corresponds to pre-conscious processing, in which all possible outputs exist in superposition. Pre-conscious quantum superposition continues until threshold for Penrose objective reduction is reached. Objective reduction (OR) – a discrete event – then occurs, actin gel dissolves, and post-OR tubulin states (chosen non-computably) proceed by classical microtubule automata to regulate synapses and other neural membrane activities (in the liquid phase of cytoplasm). The events are proposed to be conscious (to have qualia or experience) for reasons that relate to a merger of modern physics and philosophical pan-experientialism (Hameroff 1998b). A sequence of such events gives rise to a stream of consciousness.
6. Probabilities and possibilities for pre-conscious quantum superpositions are influenced by biological feedback including attachments of microtubule-associated proteins, which by modifying selected tubulins in the classical phase can tune and "orchestrate" quantum oscillations. We thus term the self-tuning OR process in microtubules "orchestrated objective reduction" or "Orch OR".
7. Orch OR events may be of variable intensity and duration of pre-conscious processing. Calculating from $E = \hbar/T$, for a pre-conscious processing time of $T = 25$ msec (corresponding to intervals with oscillations of 40 Hz), E is roughly the superposition of 2×10^{10} tubulins. For $T = 100$ msec (cor-

responding to alpha frequency EEG), E would involve 5 x 10^9 tubulins. For T = 500 msec (shown by Libet et al. 1979, as a typical pre-conscious processing time for low intensity stimuli), E is equivalent to 10^9 tubulins. Thus 2 x 10^{10} tubulins maintained in isolated quantum coherent superposition for 25 msec (or 5 x 10^9 tubulins for 100 msec, or 10^9 tubulins for 500 msec, etc.) will self-collapse according to the Orch OR model and elicit a conscious event. The interval between conscious events (or the duration of each conscious event if the pre-conscious processing time is included) varies inversely with the number of tubulins in superposition and intensity of the experience.

8. Each brain neuron is estimated to contain about 10^7 tubulins (Yu & Bass 1994). If, say, 10% of each neuron's tubulins became coherent, then Orch OR of tubulins within roughly 20,000 (gap-junction connected) neurons would be required for a 25 msec conscious event, 5,000 neurons for a 100 msec event, or 1,000 neurons for a 500 msec event, etc.
9. Each instantaneous Orch OR event binds superposed information encoded in microtubules throughout the hyperneuron whose net displacement reaches threshold at a particular moment: a variety of different modes of information is thus bound into a "now" event. As quantum state reductions are irreversible in time, cascades of Orch OR events present a forward flow of time and stream of consciousness. As OR events are actually separations and reanneallings in fundamental spacetime geometry (in which it is presumed, fundamental proto-conscious qualia reside), Orch OR events access and select particular configurations of qualia.
10. The post-OR states are chosen non-computably due to Platonic influences embedded in the Planck scale structure of spacetime geometry. Volitional choices thus involve deterministic pre-conscious processes influenced at the instant of collapse by Planck scale information; consequently in the Orch OR scheme volitional acts are neither completely deterministic, nor are they random. We suggest this can account for volition, choice and the experience of free will.

5. Conclusion

Consciousness is an exceedingly tricky problem. We have presented a model of quantum computation in cerebral cortex that addresses the enigmatic features of consciousness, such as qualia and free will, and have reconciled these events with neuroscientific details – gap junctions, dendritic processing, and metabotropic receptors – frequently ignored in conventional computation models.

References

Arieli, A., Sterkin, A., Grinvald, A., & Aertsen, A. (1996). Dynamics of ongoing activity: Explanation of the large variability in evoked cortical responses. *Science, 273*, 1868–1874.

Baars, B. (1988). *A cognitive theory of consciousness.* Cambridge, UK: Cambridge Press.

Benioff, P. (1982). Quantum mechanical Hamiltonian models of Turing Machines. *J. Stat. Phys., 29*, 515–546.

Bohm, D., & Hiley B. J. (1993). *The Undivided Universe.* New York, Routledge.

Burchinskaya, L. F. (1981). Neuronal composition and interneuronal connection of area 5 in the cat parietal association cortex. *Neurosci. Behav. Physiol., 11*, 413–420.

Buzsáki, G., & Kandel, A. (1998). Somadendritic backpropagation of action potentials in cortical pyramidal cells of the awake rat. *J. Neurophysiol., 79*, 1587–1591.

Chalmers, D. J., (1996). *The conscious mind – In search of a fundamental theory.* New York, Oxford University Press.

Conrad, M. (1994). Amplification of superpositional effects through electronic-conformational interactions. *Chaos, Solitons & Fractals, 4*, 423–438

Crick, F., & Koch, C., (1990). Towards a neurobiological theory of consciousness. *Seminars Neurosci., 2*, 263–275.

Crook, S. M., Ermentrout, G. B., & Bower, J. M. (1998). Dendritic and synaptic effects in systems of coupled cortical oscillators. *J. Comput. Neurosci., 5*, 315–329.

Dayhoff, J., Hameroff, S., Lahoz-Beltra, R., & Swenberg, C. E. (1994). Cytoskeletal involvement in neuronal learning: A review. *Eur. Biophysics J., 23*, 79–83.

De Valois, R. L., & De Valois, K. K. (1988). *Spatial Vision.* New York, Oxford University Press.

De Zeeuw, C. I., Hertzberg, E. L., & Mugnaini, E. (1995). The dendritic lamellar body: A new neuronal organelle putatively associated with dendrodendritic gap junctions. *J. Neurosci., 15*, 1587–1604.

Desmedt, J. D., & Tomberg, C. (1994). Transient phase-locking of 40 Hz electrical oscillations in prefrontal and parietal human cortex reflects the process of conscious somatic perception. *Neurosci. Letts., 168*, 126–129.

Deutsch, D. (1985). Quantum theory, the Church-Turing principle and the universal quantum computer. *Proc. Royal Soc. (London), A400*, 97–117.

Di¢si, L. 1989. Models for universal reduction of macroscopic quantum fluctuations. *Phys. Rev. A., 40*, 1165–1174.

Eccles, J. C. (1992). Evolution of consciousness. *Proc. Natl. Acad. Sci., 89*, 7320–7324.

Everett, H. (1957). Relative state formulation of quantum mechanics. In J. A. Wheeler & W. H. Zurek (Eds.), *Quantum Theory and Measurement.* Princeton University Press, 1983; originally in *Rev. Mod. Physics., 29*, 454–462.

Ferster, D. (1996). Is neural noise just a nuisance? *Science, 272*, 1812.

Feynman, R. P. (1986). Quantum mechanical computers. *Found. Physics, 16*, 507–531.
Franks, N. P., & Lieb, W. R. (1982). Molecular mechanisms of general anesthesia. *Nature, 316*, 349–351.
Fröhlich, H. (1968). Long-range coherence and energy storage in biological systems. *Int. J. Quantum Chem., 2*, 641–649.
Fröhlich, H. (1970). Long range coherence and the actions of enzymes. *Nature, 228*, 1093.
Fröhlich, H. (1975). The extraordinary dielectric properties of biological materials and the action of enzymes. *Proc. Natl. Acad. Sci., 72*, 4211–4215.
Fukuda, T., & Kosaka, T. (2000). Gap junctions linking the dendritic network of GABAergic interneurons in the hippocampus. *J. Neurosci., 20*, 1519–1528.
Galarreta, M., & Hestrin, S. (1999). A network of fast-spiking cells in the neocortex connected by electrical synapses. *Nature, 402*, 72–75.
Ghirardi, G. C., Grassi, R., & Rimini, A. (1990). Continuous-spontaneous reduction model involving gravity. *Phys. Rev. A., 42*, 1057–1064.
Ghirardi, G. C., Rimini, A., & Weber, T. (1986). Unified dynamics for microscopic and macroscopic systems. *Phys. Rev. D., 34*, 470.
Gibson, J. R., Beierlein, M., & Connors, B. W. (1999). Two networks of electrically coupled inhibitory neurons in neocortex. *Nature, 402*, 75–79.
Glanz, J. (1997). Force-carrying web pervades living cell. *Science, 276*, 678–679.
Gray, J. (1998). In S. Hameroff, A. Kaszniak, & A. Scott (Eds.), *Toward a Science of Consciousness II – The Second Tucson Discussions and Debates*. MIT Press.
Hagan, S., Hameroff, S. R., & Tuszanski, J. A. (2001). Quantum computation in brain microtubules: Decoherence and biological feasibility. *Phys. Rev. E., 65*, 061901. http://xxx.lanl.gov/abs/quant-ph/0005025
Hameroff S. R., & Watt R. C. (1982) .Information processing in microtubules. *J. Theoret. Biol., 98*, 549–561.
Hameroff, S. (1994). Quantum coherence in microtubules: a neural basis for emergent consciousness? *J. Consciousness Stud., 1*, 91–118.
Hameroff, S. (1998a). Quantum computation in brain microtubules? The Penrose-Hameroff "Orch OR" model of consciousness. *Philos. Trans. R. Soc. London Ser. A, 356*, 1869–1896. http://www.consciousness.arizona.edu/hameroff/royal2.html
Hameroff, S. (1998b). "Funda-mentality": is the conscious mind subtly linked to a basic level of the universe? *Trends Cog. Sci., 2*, 119–127.
Hameroff, S. R., & Penrose, R., (1996a). Orchestrated reduction of quantum coherence in brain microtubules: A model for consciousness. In S. Hameroff, A. Kaszniak, & A. Scott (Eds.), *Toward a Science of Consciousness The First Tucson Discussions and Debates* (pp. 507–540). MIT Press. Also published in *Mathematics and Computers in Simulation* (1996). *40*, 453–480. http://www.consciousness.arizona.edu/hameroff/or.html
Hameroff, S. R., & Penrose, R. (1996b). Conscious events as orchestrated space-time selections. *Journal of Consciousness Studies, 3*, 36–53. http://www.u.arizona.edu/~hameroff/penrose2
Isaacson, J. S., & Strowbridge, B. W. (1998). Olfactory reciprocal synapses: dendritic signaling in the CNS. *Neuron, 20*, 749–761.
Jackson, F. (1994). Finding the mind in the natural world In R. Casati, B. Smith & S. White (Eds.), *Philosophy and the Cognitive Sciences*. Vienna: Holder-Pichler-Tempsky.

Jasper, H., & Komaya, I. (1968). Amino acids released from the cortical surface in cats following stimulation of the mesial thalamus and midbrain reticular formation. *Electroencephalog. Clin. Neurophysiol., 24*, 292.

Jibu, M., Pribram, K. H., & Yasue, K. (1996). From conscious experience to memory storage and retrieval: The role of quantum brain dynamics and Boson condensation of evanescent phoyons. *Int. J. Modern Physics B, 10*, 1735–1754.

Jibu, M., Hagan, S., Hameroff, S. R., Pribram, K. H., & Yasue, K. (1994). Quantum optical coherence in cytoskeletal microtubules: Implications for brain function. *BioSystems, 32*, 195–209.

Jibu, M., & Yasue, K. (1995). *Quantum brain dynamics: An introduction.* Amsterdam: John Benjamins.

John, E. R., Tang, Y., Brill, A. B., Young, R., & Ono, K. (1986). Double layered metabolic maps of memory. *Science, 233*, 1167–1175.

Johnson, G. V. W., & Jope, R. S. (1992). The role of microtubule-associated protein 2 (MAP-2) in neuronal growth, plasticity, and degeneration. *J. Neurosci. Res., 33*, 505–512.

Joliot M., Ribary, U., & Llinas, R. (1994). Human oscillatory brain activity near 40 Hz coexists with cognitive temporal binding. *Proc. Natl. Acad. Sci. USA, 91*, 11748–11751.

Károlyházy, F., Frenkel, A., & Lukacs, B. (1986). On the possible role of gravity on the reduction of the wave function. In R. Penrose & C. J. Isham (Eds.), *Quantum Concepts in Space and Time.* Oxford University Press.

Karplus, M., & McCammon, J. A. (1983). Dynamics of proteins: elements and function. *Ann. Rev. Biochem., 52*, 263–300.

Larkman, A. U. (1991). Dendritic morphology of pyramidal neurones of the visual cortex of the rat: I. Branching patterns. *J. Comp. Neurol., 306*, 307–319.

Libet, B., Wright, E. W. Jr., Feinstein, B., & Pearl, D. K. (1979). Subjective referral of the timing for a conscious sensory experience. *Brain, 102*, 193–224.

Logothetis, N. K. (1999). Vision: a window on consciousness. *Sci. Am., 281*, 69–75.

Maniotis, A. J., Bojanowski, K., & Ingber, D. E. (1997a). Mechanical continuity and reversible chromosome disassembly within intact genomes removed from living cells. *J. Cell. Biochem., 65*, 114–130.

Maniotis A. J., Chen C. S., & Ingber D. I. (1997b). Demonstration of mechanical connections between integrins, cytoskeletal filaments, and nucleoplasm that stabilize nuclear structure. *Proc. Natl. Acad. Sci. USA, 94*, 849–854.

Molinengo, L., & Ghi, P. (1997). Behavioral and neurochemical effects induced by subchronic l-deprenyl administration. *Pharmacol. Biochem. Behav., 58*, 649–655.

Moravec, H. (1988). *Mind children: The future of robot and human intelligence.* Cambridge, MA: Harvard University Press.

Nadarajah, B., & Parnavelas, J. G. (1999). Gap junction-mediated communication in the developing and adult cerebral cortex. *Novartis Found. Symp., 219*, 157–170.

Nagel, T. (1974). What is it like to be a bat? *Philos. Rev., 83*, 435–450.

Onn, S. P., & Grace, A. A. (2000). Amphetamine withdrawal alters bistable states and cellular coupling in rat prefrontal cortex and nucleus accumbens neurons recorded in vivo. *J. Neurosci., 20*, 2332–2345.

Pearle, P. (1989). Combining stochastic dynamical state vector reduction with spontaneous localization. *Phys. Rev. D., 13*, 857–868.

Pearle, P., & Squires, E. (1994). Bound-state excitation, nucleon decay experiments and models of wave-function collapse. *Phys. Rev. Letts, 73*, 1–5.
Penrose, O., & Onsager, L. (1956). Bose-Einstein condensation and liquid helium. *Phys. Rev., 104*, 576–584.
Penrose, R., & Hameroff, S. R. (1995). What gaps? Reply to Grush and Churchland. *J. Conscious. Stud., 2*, 98–112. http://www.consciousness.arizona.edu/hameroff/gap2.html
Penrose, R. (1971). In E. A. Bastin (Ed.), *Quantum Theory and Beyond*. Cambridge, UK: Cambridge University Press.
Penrose, R. (1989). *The emperor's new mind*. Oxford University Press.
Penrose, R. (1994). *Shadows of the mind: A search for the missing science of consciousness*. Oxford University Press.
Penrose, R. (1996). On gravity's role in quantum state reduction. *Gen. Relativity Gravitation, 28*, 581–600.
Perry, E., Walker, M., Grace, J., & Perry, R. (1999). Acetylcholine in mind: a neurotransmitter correlate of consciousness? *Trends Neurosci., 22*, 273–280.
Porter, M. (2001). Topological quantum computation/error correction in microtubules. http://www.consciousness.arizona.edu/hameroff/topqcomp.htm
Pribram, K. H. (1991). *Brain and Perception*. New Jersey: Lawrence Erlbaum.
Ramón y Cajal, S. (1909). *Histologie du Systèm Nerueux de L'homme & des Vértébrates.*
Rasmussen, S., Karampurwala, H., Vaidyanath, R., Jensen, K. S., & Hameroff, S. (1990). Computational connectionism within neurons: A model of cytoskeletal automata subserving neural networks. *Physica D, 42*, 428–449.
Roitberg, A., Gerber, R. B., Elber, R., & Ratner, M. A. (1995). Anharmonic wave functions of proteins: Quantum self-consistent field calculations of BPTI. *Science, 268*, 1319–1322.
Sánchez, C., Díaz-Nido, J., & Avila, J. (2000). Phosphorylation of microtubule-associated protein 2 (MAP2) and its relevance for the regulation of the neuronal cytoskeleton function. *Prog. Neurobiol., 61*, 133–168.
Sarter, M., & Bruno, J. P. (1999). Cortical cholinergic inputs mediating arousal, attentional processing and dreaming: Differential afferent regulation of the basal forebrain by telencephalic and brainstem afferents. *Neuroscienc, 95*, 933–952.
Sassoè-Pognetto, M., & Ottersen, O. P. (2000). Organization of ionotropic glutamate receptors at dendrodendritic synapses in the rat olfactory bulb. *J. Neurosci., 20*, 2192–2201.
Schrodinger, E. (1935). Die gegenwarten situation in der quantenmechanik. *Naturwissenschaften, 23*, 807–812, 823–828, 844–849. (Translation by J. T. Trimmer (1980) in *Proc. Amer. Phil. Soc., 124*, 323–338.)
Schwindt, P. C., & Crill, W. E. (1998). Synaptically evoked dendritic action potentials in rat neocortical pyramidal neurons. *J. Neurophysiol., 79*, 2432–2446.
Seife, C. (2000). Cold numbers unmake quantum mind. *Science, 287*, 791.
Shimony, A., (1993). *Search for a Naturalistic World View – Volume II. Natural Science and Metaphysics*. Cambridge, UK: Cambridge University Press.
Siegel, G. J., Agranoff, B. W., Albers, R. W., Fisher, S. K., & Uhler, M. D. (1998). *Basic Neurochemistry: Molecular, Cellular and Medical Aspects*. Lippincott-Raven.

Singer, W. (1999). Neuronal synchrony: a versatile code for the definition of relations? *Neuron, 24*, 111–125.

Singer, W., Gray, C., Engel, A., Konig, P., Artola, A., & Brocher, S. (1990) Formation of cortical cell assemblies. *Cold Spring Harbor Symposia on Quantitative Biology, 55*, 939–952.

Smolin, L., (1997). *Life of the Cosmos*. N.Y.: Oxford Press.

Sourdet, V., & Debanne, D. (1999). The role of dendritic filtering in associative long-term synaptic plasticity. *Learning and Memory, 6*, 422–447.

Spinoza, B. (1677). In J. van Vloten & J. P. N. Land (Eds.), *Ethica in Opera quotque reperta sunt* (3rd edition). Netherlands, Den Haag.

Stockley, E. W., Cole, H. M., Brown, A. D., & Wheal, H. V. (1993). A system for quantitative morphological measurement and electronic modelling of neurons: three-dimensional reconstruction. *J. Neurosci. Meth., 47*, 39–51.

Stuart, G., Schiller, J., & Sakmann, B. (1997). Action potential initiation and propagation in rat neocortical pyramidal neurons. *J. Physiol., 505*, 617–632.

Tall, E. G., Spector, I., Pentyala, S. N., Bitter, I., & Rebecchi, M. J. (2000). Dynamics of phosphatidylinositol 4,5-bisphosphate in actin-rich structures. *Current Biol., 10*, 743–746.

Tamas, G., Buhl, E. H., Lörincz, A., & Somogyi, P. (2000). Proximally targeted GABAergic synapses and gap junctions synchronize cortical interneurons. *Nat. Neurosci., 3*, 366–371.

Tegmark, M. (2000). The importance of quantum decoherence in brain processes. *Phys. Rev. E, 61*, 4194–4206.

Tejada, J., Garg, A., Gider, S., Awschalom, D. D., DiVincenzo, D. P., & Loss, D. (1996). Does macroscopic quantum coherence occur in ferritin? *Science, 272*, 424–426.

Vernon, G. G., & Woolley, D. M. (1995). The propagation of a zone of activation along groups of flagellar doublet microtubules. *Exp. Cell Res., 220*, 482–494.

Voet, D., & Voet, J. G. (1995). *Biochemistry* (2nd edition). New York: Wiley.

von der Heydt, R., Peterhans, E., & Duersteler (1992). Periodic-pattern-selective cells in monkey visual cortex. *J. Neurosci., 12*, 1416–1434.

Von Neumann, J. (1966). In A. W. Burks (Ed.), *Theory of Self-Reproducing Automata*. Urbana: University of Illinois Press.

Wheeler, J. A. (1990). Information, physics, quantum: The search for links. In W. Zurek (Ed.), *Complexity, Entropy, and the Physics of Information*. Addison-Wesley.

Whitehead, A. N. (1929). *Process and Reality*. New York: Macmillan.

Woolf, N. J. (1996). Global and serial neurons form a hierarchically-arranged interface proposed to underlie learning and cognition. *Neuroscience, 74*, 625–651.

Woolf, N. J. (1997). A possible role for cholinergic neurons of the basal forebrain and pontomesencephalon in consciousness. *Conscious. Cog., 6*, 574–596.

Woolf, N. J. (1998). A structural basis for memory storage in mammals. *Prog. Neurobiol., 55*, 59–77.

Woolf, N. J. (1999a). Cholinergic correlates of consciousness: from mind to molecules. *Trends Neurosci., 22*, 540–541.

Woolf, N. J. (1999b). Dendritic encoding: An alternative to temporal synaptic coding of conscious experience. *Conscious. Cog., 8*, 574–596.

Woolf, N. J., & Hameroff, S. R. (2001). A quantum approach to visual consciousness. *Trends in Cog. Sci., 5*, 472–478.

Woolf, N. J., Zinnerman, M. D., & Johnson, G. V. W. (1999). Hippocampal microtubule-associated protein-2 alterations with contextual memory. *Brain Res., 821*, 241–249.

Yu, W., & Baas, P. W. (1994). Changes in microtubule number and length during axon differentiation. *J. Neurosci., 14*, 2818–2829.

CHAPTER 11

On quantum theories of the mind

Alwyn Scott
University of Arizona

Introduction

Over the past few years, discussions of the roles that quantum mechanics might or might not play in the theory of consciousness have become increasingly sharp. On one side of this debate stand conventional neuroscientists who assert that brain science must look to the neuron for understanding, and on the other side are certain physicists, suggesting that the rules of quantum theory might influence the dynamics of mind. Often it has seemed that proponents of these two views are talking past each other, with neither side listening very carefully. My position on this debate is rather uncomfortable. While appreciating the emphasis placed on neural networks by the community of neuroscientists, I share the motivating conviction of the quantum enthusiasts that mind is something more than mere neurodynamics. The aim of this paper is to indicate a middle way between the vagueness of theories for quantum consciousness and the sterility of the neural network approach. How is this possible? Both sides of the current debate in my opinion give insufficient consideration to the explanatory power of classical nonlinear theory. Properly appreciated, the ramifications of intricate nonlinear systems obviate the need for turning – in desperation, one may say – to quantum theory as a source of the mysteries of mind. At the same stroke, it rescues neural network theory from the shallows of a narrowly conceived functionalism. Throughout the discussion, effort is made to remain grounded in experimental reality. Theoretical speculation is avoided in favor of statements that are based upon empirical observations.

Classical nonlinear dynamics

In works that argue for the necessity of quantum effects in the dynamics of mind, one often comes across statements like the following (Stapp 1993): Nothing in classical physics can create something that is essentially more than an aggregation of its parts. Such a statement is evidence of a belief in the constructive power of reductionism that is at variance with a considerable amount of theoretical and experimental knowledge accumulated over the past three decades in the field of nonlinear science. A differing view has been offered by Philip Anderson, a Nobel Laureate in condensed matter physics, who argued that (Anderson 1972):

> The reductionist hypothesis does not by any means imply a 'constructionist' one: The ability to reduce everything to simple fundamental laws does not imply the ability to start with those laws and construct the universe. In fact the more the elementary particle physicists tell us about the nature of the fundamental laws, the less relevance they seem to have to the very real problems of the rest of science, much less to those of society.

The reason that the constructionist hypothesis breaks down in the realm of biology is that biological dynamics is both very intricate and also nonlinear. What does it mean to say that the dynamics of a system are nonlinear? In its deepest sense, this is a statement about the nature of causality. Suppose that a series of experiments on a certain system has shown that cause C1 gives rise to effect E1; thus

$$C1 \rightarrow E1$$

and similarly

$$C2 \rightarrow E2$$

expresses the relationship between cause C2 and effect E2, where the arrow indicates the action of the system being studied. Mathematicians say a system is linear if

$$C1 + C2 \rightarrow E12 \text{ equals } E1 + E2$$

which tells us that two causes acting together (CI + C2) Ieads to an effect (E12) that is the sum of the two individual effects (EI + E2). The whole is equal to the sum of its parts. For a nonlinear system, on the other hand, the whole is not equal to the sum of its parts. In symbols,

$$C1 + C2 \rightarrow E12 \text{ does NOT equal } E1 + E2$$

which tells us that the effect of two causes acting together is not the sum of the individual effects. Thus the key feature of nonlinear dynamics is to create something (E12) that differs from an aggregation of its parts (EI + E2). Can this creation be "essentially" different? To those who would answer that a classical nonlinear creation cannot be essentially different from an aggregation of its parts, one can pose the following questions: Is not liquid water essentially different from gaseous hydrogen and oxygen? Is not a tornado essentially different from the equations of aerodynamics? Is not a bacterium essentially different from the oily scum of the late Hadean oceans, out of which it emerged some three or four billion years ago? Is not a reader of this chapter essentially different from her first ancestral bacterium? These are not idle questions. Those who deny of the explanatory power of classical science should look about themselves and see the natural world as it really is: a pandemonium of interwoven nonlinearities, each biting its own tail like the uroboros of antiquity (Scott 1995). To gain an appreciation for the vigor and variety of current research in classical nonlinear dynamics, one should page through some recent issues of Physica D, Nonlinearity, or Physical Review, particularly sections B and E.

The linearity of quantum theory

In the insightful formulation of the Austrian physicist Erwin Schroedinger (Schroedinger 1926), components of a quantum wave function are calculated as solutions of a linear wave equation, appropriately called Schroedinger's equation. These components are then simply added together in a linear fashion to obtain a global wave function (or wave packet), which is interpreted as a probability density for finding the system in one or another of its characteristic states (eigenstates). Thus quantum theory is linear in Schroedinger's representation. The property of linearity implies that quantum wave packets lack the coherence that is a signature of classical nonlinear systems. As a well studied example, consider a soliton, which from the perspective of classical nonlinear dynamics is a globally coherent lump of energy, moving without change of shape or speed, even after colliding with other such lumps. Classically, this soliton is a well defined thing. From a quantum perspective, however, the linear wave function describing the lump spreads out because its various components move at different speeds (Scott 1999). The lump of energy is said to disperse, an effect that follows directly from the linearity of Schroedinger's equation.

Occasionally it is asserted that quantum systems typically display globally coherent, emergent properties, whereas nonlinear classical systems do not

(Zohar 1996), but such claims are not in accord with established research results in nonlinear dynamics. An example of the difficulties that quantum theory encounters in attempting to represent interesting behaviors of classical nonlinear systems is the local mode effect in small molecules. Spectral measurements by physical chemists of the carbon-hydrogen stretching oscillations in benzene, for example, show unmistakable evidence of vibrational energy that is localized at or near a single carbon-hydrogen site (Henry 1976), but early experimental manuscripts reporting local modes were actually rejected as being incompatible with quantum theory (Henry 2001)! Although local mode behavior arises naturally in nonlinear classical systems, it was argued from the perspective of quantum theory that the vibrational energy must be spread out over all the available sites; thus the experiments were judged to be wrong.

It is now understood from a variety of theoretical studies and experimental measurements that quantum theory manages to mimic the local mode behavior of classical nonlinear theory by generating eigenstates with almost the same energies (quasi-degenerate states) so the linear process of quantum dispersion proceeds slowly (Bernstein, Eilbeck, & Scott 1990). The take-home message is this: Stable and globally coherent states that arise naturally in classical nonlinear systems are aped with difficulty by the wave packets of quantum mechanics.

Note finally that quantum theory does not give wrong answers in the realm of classical dynamics, it merely offers a more difficult way to solve Newton's second law of motion (force equals mass times acceleration) that yields no additional information. Think of a tennis ball that Martina Hingis has just belted into the far corner of her opponent's court. If a quantum mechanical wave function is used to describe its motion, the spreading out (or dispersion) of this wave function causes an error of about .000 000 000 000 01% in calculating its position (Scott 1995). There is no way that such a small correction could influence a realistic calculation of the ball's trajectory.

Quantum biochemistry

We now briefly examine the roles played by quantum theory in the science of biochemistry. For proponents of a quantum theory of the mind, this is a key area of investigation. Why? Because biochemistry is not far removed from atomic physics (where quantum theory is necessary), and the effects of biochemically active drugs on states of consciousness have been amply established.

In a first approximation, biochemists view a protein or a nucleic acid as a collection of classical atoms that are interacting through nonlinear forces (McCammon & Harvey 1987). To what extent do they use quantum theory? Biochemists would reply that they use quantum theory where it is appropriate and not when it isn't.

Quantum theory is needed, for example, to calculate the nonlinear forces between atoms.

The key idea behind this calculation was provided in 1927 by J. Robert Oppenheimer, who had just earned a doctorate in physics under Max Born at Goettingen. The nucleus of an atom, young Oppenheimer pointed out, is much heavier than an electron, so it moves much more slowly. Assuming the atomic centers to be stationary greatly simplifies a quantum mechanical calculation of the electronic energy, from which the interatomic forces are readily derived. Basing his analysis upon Schroedinger's recent formulation of quantum mechanics, Born developed Oppenheimer's off-hand suggestion into a rigorous technique that remains the standard method for calculating interatomic forces to this day (Born & Oppenheimer 1927).

For our purposes, this approach to the theory of molecular dynamics is important because it provides reliable estimates of the errors involved in neglecting quantum effects. Having estimated the interatomic forces of a molecule from the Born-Oppenheimer method or from direct experimental measurements or through a judicious blend of the two approaches , one can consider whether or not it is appropriate to use quantum mechanics to calculate some aspect of the atomic motion. Five aspects are of particular interest.

1. Interatomic vibrations. Roughly speaking, two neighboring atoms in a molecule will act as if connected by a nonlinear spring, and the oscillations of the resulting mass spring system will be governed by the rules of quantum mechanics, which restrict the vibrational energy to certain discrete levels. It is through measurements of these levels that the experimentalist helps the theoretician refine parameters in the nonlinear interatomic force fields.
2. Molecular dynamics. In calculating the large scale motions of atoms or molecules, it is important to consider the quantum mechanical wave length of a moving atom because this length indicates the scale of distance over which an atom or molecule can be localized. Since this localization length is inversely proportional to the mass of a moving object and the mass is unity for a hydrogen atom, it turns out that a hydrogen atom can be localized to a region that is about equal to its size. For a molecule of oxygen, on

the other hand, the mass is 32 times larger, so quantum theory allows it to be localized to a region that is a few percent of its size.

To study the way that two hydrogen molecules and an oxygen molecule combine to form two molecules of water, therefore, the biochemist uses Newton's second law, where the interatomic forces are obtained from the Born-Oppenheimer procedure. Several numerical codes are currently available to perform such classical nonlinear computations.

3. Polarons. First suggested by the Soviet physicist Lev Landau in 1933, the polaron is a means for transporting localized electric charge or vibrational energy through a large molecule or crystal by locally distorting the lattice (Scott 1992, 1999). (The basic idea can be appreciated by thinking of how a marble works its way through a plate of spaghetti.) From the perspective of nonlinear classical dynamics, the polaron phenomenon is both well understood and stable. To the extent that a quantum description is appropriate, the classical lump of energy will disperse because each component of its linear wave packet travels at a different speed. Again, the effect of a quantum correction to the extent that it makes any difference at all is to degrade the global coherence of the classical polaron.
4. Solitons in microtubules. One of the more interesting components in a living cell is the protein tubulin, which forms many useful structures; including flagella and cilia (propelling bacteria in their search for food), (centrioles helping eukaryotic cells divide and the cytoskeleton which), (defines the cell structure and transports nutrients). Comprising a vast array of microtubules, the cytoskeleton may also play a role in the cell's internal information processing (Sherrington 1951). If so, such internal computations might employ localized solutions (solitons) of classical nonlinear equations to transport bits of information (Hameroff & Watt 1982). Are quantum effects to be anticipated in the dynamics of microtubule solitons? The above considerations suggest not, for the mass of a signiflcant portion of an individual tubulin molecule runs easily into the tens of thousands, indicating that quantum delocalization is limited to a very small fraction of an atomic diameter. Even if quantum phenomena were of importance, however, the primary effect would again be to degrade the global coherence of the classical nonlinear phenomena.
5. Macroscopic quantum states. Of greatest interest to those who assert the relevance of quantum theory for brain science is the possibility that elementary biological excitations (membrane oscillations, internal modes of protein vibration, electric charge, etc.) may condense into quantum states of macroscopic dimensions (Zohar 1996). Such phenomena are indeed ob-

served in the real world of condensed matter physics: the superconductivity of certain metals and oxide crystals, the superfluidity of liquid helium, and the coherent light field of a Laser oscillator being dramatic examples. When they occur, macroscopic quantum effects are rather easy to recognize; for example, the resistance to direct current of a superconductor falls exactly to zero at low temperature as does the viscosity of a superfluid, and the light output from a Laser has striking properties that have amazed and amused us all in museums of science. Since it would be of great interest to discover similar phenomena in the cytoskeleton or synapse of a nerve cell, it is well to keep the experimentalists looking in this direction (Hameroff 1994; Eccles 1994; Penrose 1994). One should note, however, that effective models of macroscopic quantum phenomena are provided by classical nonlinear wave equations (Ginzburg-Landau equations for superconductivity and superfluidity, and Maxwell's equations augmented by nonlinear rate equations for the Laser), as solutions of these classical equations adequately describe the experimentally properties of the condensed states.

The Hodgkin-Huxley equations

One of the great success stories in mathematical biophysics began in 1952 with the publication by Alan Hodgkin and Andrew Huxley of a detailed theory for the nonlinear propagation of an action potential along a nerve fiber (Scott 2002; Hodgkin & Huxley 1952). Their formulation is now accepted across the globe as the fundamental equations of electrophysiology because it satisfies the three necessary requirements of a successful theory: (i) The Hodgkin-Huxley theory contains no arbitrarily adjustable or unmeasurable parameters, (ii) It has withstood the assaults of hundreds of critical experimental tests, and (iii) It successfully predicts an impressive variety of interesting and unexpected phenomena (Scott 1975). Without doubt, this theory is "right" in the sense that the word is ordinarily used among scientists. Since the Hodgkin-Huxley theory is entirely classical, we might consider whether it can be viewed as an approximation to some more exact quantum formulation.

If this were the possible, errors in measurements of the position of a propagating nerve impulse would be about .000 000 1% (Scott 1995).

While not as impressive as our previous error estimate for Martina's tennis ball, such a small quantum correction would be undetectable by an electrophysiologist. But there is a deeper reason why quantum theory should not be applied to the description of a propagating nerve impulse: It is not possible in principle. Why is this so?

Schroedinger's equation, upon which quantum theory is based, is necessarily constructed from a classical theory that conserves energy, but nerve impulse solutions of the Hodgkin-Huxley equations do not conserve energy. Instead, they balance the rate at which electrostatic energy is released from electric fields in the nerve membrane to the power that is consumed in circulating currents. Such dynamic processes called nonlinear diffusion are of vital importance throughout the realm of biology. If the concept of nonlinear diffusion seems difficult to grasp, think of a lighted candle.

The heat of the flame releases a stream of energy laden vapor from the wax, which keeps the flame hot. The flame is a stable dynamic entity that balances power instead of conserving energy. As such, it is an appropriate metaphor for a nerve impulse. Since a nerve impulse does not conserve energy, there is no way to con-struct a Schroedinger equation for its quantum mechanical description. The classical Hodgkin-Huxley impulse of electrophysiology does not have a corresponding quantum wave packet. We shall meet a similar situation in the following section.

Schroedinger's cat

During discussions of the applications of quantum theory to biological problems, one often encounters Schroedinger's cat (Schroedinger 1935). In this famous thought experiment (I hope no one has ever tried it!), a cat is enclosed in a sealed box for an hour, during which time a radioactive source has a probability of 50% for emitting an alpha-particle. If an alpha-particle is emitted, a mechanism is triggered that breaks a vial of potassium cyanide, killing the poor cat. If an alpha-particle is not emitted from the radioactive source, the cat lives.

Blindly applied, quantum theory suggests to some that the cat is neither alive nor dead just before the box is opened. Rather it is represented by a mixture of the quantum mechanical wave functions of a living cat with that of a dead cat. Upon opening the box, it is seriously suggested, this mixed wave function "collapses" into one of two possibilities: a live cat or a dead cat.

What does this thought experiment tell us about the relevance of quantum theory for descriptions of biological reality? Let us be clear on one point straight-away. Erwin Schroedinger – who knew something about both quantum theory and biology – intended this as no more than a "burlesque example" of how quantum theory might be misused in the realm of biology. (Albert

Einstein's favorite example was a wave function representing all of the gunpowder in Germany.) Anticipating the possibility that some may disagree with Schroedinger and Einstein, let us assume for the purposes of argument that it is possible to construct a quantum mechanical wave packet that mixes the wave function of a live cat with that of a dead cat in equal proportions. Defenders of this position that the cat is neither alive nor dead just before the box is opened should consider the following objections.

1. Since a cat does not conserve energy, it is not possible to construct the Schroedinger equation for which the wave function in question is presumed to be a solution.
2. Even if it were possible to find such a wave function, a conservative estimate suggests that a time very much longer than the age of the universe would be required for the cat's wave function to rotate from being dead to being alive (Scott 1995). If the wave function components were to remain coherent, therefore, this implies that a quantum mechanical cat would be either dead or alive but not both.
3. The time scale over which the components of a biological wave function could possibly remain coherent in the face of the thermal and structural disorder in a normal brain is very short. Quantitatively, this coherence time is some twenty orders of magnitude below typical times involved in neural dynamics (Tegmark 2000).
4. Upon opening the lid and finding the cat dead, the conscious observer could determine when it had died by measuring the relative amounts of oxygen and carbon dioxide remaining in the box.
5. Cat lovers might prefer to do the experiment with a lighted candle, whereupon the remaining length of the taper would indicate when the flame had been snuffed out.

In the light of such objections, I suggest that we agree with Schroedinger: It is inappropriate to use a quantum wave packet to describe a living organism. How much less appropriate is it then to use this theory to describe the dynamics of a brain?

Conclusions

Current research in the science of consciousness spans a wide range of approaches, from a narrowly conceived functionalism that would deny free will and question the reality of qualia to a facile dualism that attempts to encompass the evident realities of human existence. Both of these philosophical extremes discount the deep explanatory power of classical nonlinear dynamics. The existence of this explanatory power is not a matter of hypothesis; it has been amply demonstrated over the past three decades by experimental measurements, numerical studies and theoretical analyses in many fields of science, including life sciences from molecular biology to the dynamics of evolutionary development.

In each of these sciences, studies of classical nonlinear equations lead to an unexpected (not to say bewildering) variety of natural phenomena that are unrelated to quantum theory. Considering a human mind to be organized into many functional layers of activity starting with the dynamics of membrane proteins and the interactions of nerve impulses in a single neuron, progressing upward through the functional organization of neurons into cell assemblies and of assemblies into higher order assemblies (Scott 2002; Hebb 1949; Braitenberg 1982; Palm 1982; Hopfield 1982; Freeman & Skarda 1985), to the multifaceted relationship between a brain and its cultural configuration it is difficult to put any bounds on the nature of the mental phenomena that might emerge from classical nonlinear science (Scott 1995).

All of the above is not to claim that classical nonlinear dynamics has already succeeded in explaining consciousness. Although the spectrum of behaviors that can emerge at a single layer of dynamic activity (global coherence, threshold effects, binding of organized states, and so on) is becoming understood, the study of hierarchies of such systems has barely begun, and the qualitative natures of their phenomena are largely unknown. But a better appreciation for the dynamic repertory of nonlinear science can open two doors in consciousness research. For the functionalists, such appreciation can expand their concept of what is possible in the realm of natural phenomena. Dualists, similarly, are unburdened of the need to appeal to quantum theory for doubtful explanations of what they know to be true.

References

Anderson, P. W. (1972). More is different: Broken symmetry and the nature of the hierarchical structure of science. *Science, 177*, 393–396.

Bernstein, L., Eilbeck, J. C., & Scott, A. C. (1990). The quantum theory of local modes in a coupled system of nonlinear oscillators. *Nonlinearity, 3*, 293–323.

Born, M., & Oppenheimer, J. R. (1927). Zur Quantentheorie der Molekeln. *Annalen der Physik, 84*, 57–484.

Braitenberg, V. (1982). Cell assemblies in the cerebral cortex. In R. Heim & G. Palm (Eds.), *Theoretical approaches to complex systems.* Berlin: Springer-Verlag.

Eccles, J. C. (1994). *How the self controls its brain*. Berlin: Springer-Verlag.

Freeman, W., & Skarda, C. (1985). Spatial EEG patterns, nonlinear dynamics and perception: The neo-Sherrington view. *Brain Research Review, 10*, 147–175.

Hameroff, S. R., & Watt, R. C. (1982). Information processing in micro-tubules. *Journal of Theoretical Biology, 98*, 549–561.

Hameroff, S. R. (1994). Quantum consciousness in microtubules: An intra-neuronal substrate for emergent consciousness. *Journal of Consciousness Studies, 1*, 91–118.

Hebb, D. O. (1949). *The Organization of behavior*. New York: John Wiley & Sons.

Henry, B. R. (1976). Local modes and their application to the analysis of polyatomic overtone spectra. *Journal of Physical Chemistry, 80*, 2160–2164.

Henry, B. R. (1988). Private communication.

Hodgkin, A. L., & Huxley, A. F. (1952). A quantitative description of membrane current and its application to conduction and excitation in nerve. *Journal of Physiology, 117*, 500–544.

Hopfield, J. (1982). Neural networks and physical systems with emergent collective computational abilities. *Proceedings of the National Academy of Sciences, USA, 79*, 2554–2558.

Landau, L. D. (1933). Ueber die Bewegung der Elektronen in Kristalgitter. *Zeitschrift der Sovietunion, 3*, 664–665.

McCammon, J. A., & Harvey, S. C. (1987). *Dynamics of proteins and nucleic acids.* Cambridge: Cambridge University Press.

Palm, G. (1982). *Neural assemblies: An alternative approach to artificial intelligence.* Berlin: Springer-Verlag.

Penrose, R. (1994). *Shadows of the mind: A search for the missing science of consciousness.* Oxford: Oxford University Press.

Scott, A. C. (1975). The electrophysics of a nerve fiber. *Reviews of Modern Physics, 47*, 487–533.

Scott, A. C. (1992). Davydov's soliton. *Physics Reports, 217*, 1–67.

Scott, A. C. (1995). *Stairway to the mind*. New York: Springer-Verlag (Copernicus).

Scott, A. C. (1999). *Nonlinear science: Emergence and dynamics of coherent structures*. Oxford: Oxford University Press.

Scott, A. C. (2002). *Neuroscience: A Mathematical Primer*. New York: Springer-Verlag.

Sherrington, C. S. (1951). *Man on his nature* (2nd ed.). Cambridge: Cambridge University Press.

Schroedinger, E. (1926). Quantisierung als Eigenwertproblem. *Annalen der Physik, 79*, 361–376.

Schroedinger, E. (1935). Die gegenwaertige Situation der Quantenmechanik. *Naturwissenschaften, 23*, 807–812, 823–828, 844–849.

Stapp, H. P. (1993). *Mind, matter, and quantum mechanics*. Berlin: Springer-Verlag.

Tegmark, M. (2000). The importance of quantum decoherence in brain processes. *Physical Review E, 61*, 4194–4206.

Zohar, D. (1996). Consciousness and Bose-Einstein condensates. In S. R. Hameroff, A. W. Kaszniak, & A. C. Scott (Eds.), *Toward a science of consciousness* (pp. 439–450). MIT Press.

Name index

Subject index

In the series ADVANCES IN CONSCIOUSNESS RESEARCH (AiCR) the following titles have been published thus far or are scheduled for publication:

1. GLOBUS, Gordon G.: *The Postmodern Brain.* 1995.
2. ELLIS, Ralph D.: *Questioning Consciousness. The interplay of imagery, cognition, and emotion in the human brain.* 1995.
3. JIBU, Mari and Kunio YASUE: *Quantum Brain Dynamics and Consciousness. An introduction.* 1995.
4. HARDCASTLE, Valerie Gray: *Locating Consciousness.* 1995.
5. STUBENBERG, Leopold: *Consciousness and Qualia.* 1998.
6. GENNARO, Rocco J.: *Consciousness and Self-Consciousness. A defense of the higher-order thought theory of consciousness.* 1996.
7. MAC CORMAC, Earl and Maxim I. STAMENOV (eds): *Fractals of Brain, Fractals of Mind. In search of a symmetry bond.* 1996.
8. GROSSENBACHER, Peter G. (ed.): *Finding Consciousness in the Brain. A neurocognitive approach.* 2001.
9. Ó NUALLÁIN, Seán, Paul MC KEVITT and Eoghan MAC AOGÁIN (eds): *Two Sciences of Mind. Readings in cognitive science and consciousness.* 1997.
10. NEWTON, Natika: *Foundations of Understanding.* 1996.
11. PYLKKÖ, Pauli: *The Aconceptual Mind. Heideggerian themes in holistic naturalism.* 1998.
12. STAMENOV, Maxim I. (ed.): *Language Structure, Discourse and the Access to Consciousness.* 1997.
13. VELMANS, Max (ed.): *Investigating Phenomenal Consciousness. Methodologies and Maps.* 2000.
14. SHEETS-JOHNSTONE, Maxine: *The Primacy of Movement.* 1999.
15. CHALLIS, Bradford H. and Boris M. VELICHKOVSKY (eds.): *Stratification in Cognition and Consciousness.* 1999.
16. ELLIS, Ralph D. and Natika NEWTON (eds.): *The Caldron of Consciousness. Motivation, affect and self-organization – An anthology.* 2000.
17. HUTTO, Daniel D.: *The Presence of Mind.* 1999.
18. PALMER, Gary B. and Debra J. OCCHI (eds.): *Languages of Sentiment. Cultural constructions of emotional substrates.* 1999.
19. DAUTENHAHN, Kerstin (ed.): *Human Cognition and Social Agent Technology.* 2000.
20. KUNZENDORF, Robert G. and Benjamin WALLACE (eds.): *Individual Differences in Conscious Experience.* 2000.
21. HUTTO, Daniel D.: *Beyond Physicalism.* 2000.
22. ROSSETTI, Yves and Antti REVONSUO (eds.): *Beyond Dissociation. Interaction between dissociated implicit and explicit processing.* 2000.
23. ZAHAVI, Dan (ed.): *Exploring the Self. Philosophical and psychopathological perspectives on self-experience.* 2000.
24. ROVEE-COLLIER, Carolyn, Harlene HAYNE and Michael COLOMBO: *The Development of Implicit and Explicit Memory.* 2000.
25. BACHMANN, Talis: *Microgenetic Approach to the Conscious Mind.* 2000.
26. Ó NUALLÁIN, Seán (ed.): *Spatial Cognition. Selected papers from Mind III, Annual Conference of the Cognitive Science Society of Ireland, 1998.* 2000.
27. McMILLAN, John and Grant R. GILLETT: *Consciousness and Intentionality.* 2001.

28. ZACHAR, Peter: *Psychological Concepts and Biological Psychiatry. A philosophical analysis.* 2000.
29. VAN LOOCKE, Philip (ed.): *The Physical Nature of Consciousness.* 2001.
30. BROOK, Andrew and Richard C. DeVIDI (eds.): *Self-reference and Self-awareness.* 2001.
31. RAKOVER, Sam S. and Baruch CAHLON: *Face Recognition. Cognitive and computational processes.* 2001.
32. VITIELLO, Giuseppe: *My Double Unveiled. The dissipative quantum model of the brain.* 2001.
33. YASUE, Kunio, Mari JIBU and Tarcisio DELLA SENTA (eds.): *No Matter, Never Mind. Proceedings of Toward a Science of Consciousness: Fundamental Approaches, Tokyo, 1999.* 2002.
34. FETZER, James H.(ed.): *Consciousness Evolving.* 2002.
35. Mc KEVITT, Paul, Seán Ó NUALLÁIN and Conn MULVIHILL (eds.): *Language, Vision, and Music. Selected papers from the 8th International Workshop on the Cognitive Science of Natural Language Processing, Galway, 1999.* 2002.
36. PERRY, Elaine, Heather ASHTON and Allan YOUNG (eds.): *Neurochemistry of Consciousness. Neurotransmitters in mind.* 2002.
37. PYLKKÄNEN, Paavo and Tere VADÉN (eds.): *Dimensions of Conscious Experience.* 2001.
38. SALZARULO, Piero and Gianluca FICCA (eds.): *Awakening and Sleep-Wake Cycle Across Development.* 2002.
39. BARTSCH, Renate: *Consciousness Emerging. The dynamics of perception, imagination, action, memory, thought, and language.* 2002.
40. MANDLER, George: *Consciousness Recovered. Psychological functions and origins of conscious thought.* 2002.
41. ALBERTAZZI, Liliana (ed.): *Unfolding Perceptual Continua.* 2002.
42. STAMENOV, Maxim I. and Vittorio GALLESE (eds.): *Mirror Neurons and the Evolution of Brain and Language.* 2002.
43. DEPRAZ, Natalie, Francisco VARELA and Pierre VERMERSCH.: *On Becoming Aware.* n.y.p.
44. MOORE, Simon and Mike OAKSFORD (eds.): *Emotional Cognition. From brain to behaviour.* 2002.
45. DOKIC, Jerome and Joelle PROUST: *Simulation and Knowledge of Action.* 2002.
46. MATHEAS, Michael and Phoebe SENGERS (ed.): *Narrative Intelligence.* n.y.p.
47. COOK, Norman D.: *Tone of Voice and Mind. The connections between intonation, emotion, cognition and consciousness.* 2002.
48. JIMÉNEZ, Luis: *Attention and Implicit Learning.* 2003.
49. OSAKA, Naoyuki (ed.): *Neural Basis of Consciousness.* 2003.
50. GLOBUS, Gordon G.: *Quantum Closures and Disclosures. Thinking-together post-phenomenology and quantum brain dynamics.* n.y.p.